POWER SYSTEMS AND RENEWABLE ENERGY

POWER SYSTEMS AND RENEWABLE ENERGY

DESIGN, OPERATION, AND SYSTEMS ANALYSIS

Gary D. Price

MP MOMENTUM PRESS

MOMENTUM PRESS, LLC, NEW YORK

Power Systems and Renewable Energy: Design, Operation, and Systems Analysis
Copyright © Momentum Press®, LLC, 2014.

Previous edition published by Gary D. Price in 2011. Copyright © 2011 by Gary D. Price

First published by Momentum Press®, LLC
222 East 46th Street, New York, NY 10017
www.momentumpress.net

ISBN-13: 978-1-60650-570-0 (print)
ISBN-13: 978-1-60650-571-7 (e-book)

Momentum Press Power Generation Collection

DOI: 10.5643/9781606505717

Cover design by Jonathan Pennell
Interior design by Exeter Premedia Services Private Ltd.,
Chennai, India

10 9 8 7 6 5 4 3 2 1

Printed in the United States of America

ABSTRACT

Solar and wind energy systems have flourished throughout the United States in the last few years as the public calls for reduced dependence on foreign oil. Government programs have been established to meet the public demand. Many states have passed legislation that requires electric utilities to include a portfolio of renewable energy sources in their generation mix. The resulting public demand has stimulated the growth of an industry that provides wind and solar systems, and many small businesses have grown to install these systems. Training programs and courses are now ubiquitous as the demand for designers and installers increases, and almost every educational institution offers renewable energy classes or curriculum.

The goal is to provide a resource for engineering students interested in the design and operation of solar electric, solar thermal, wind, and other renewable systems. In particular, the author found that a text that provides challenging problems and solutions that stimulate engineering thinking was necessary. While there are many good reference books on power systems and renewable energy, the objective here is to integrate the engineering basics of existing power systems with design problems and solutions using renewable energy sources.

The organization of this book begins with concepts and terminology for power and energy—the basics needed to communicate and understand the subject. Conventional power systems are briefly discussed to understand the concepts used in the integration of renewable power systems. We quickly move to the design and installation of a small residential photovoltaic system and wind generator connected to the electric utility grid. With this background, the student can begin developing ideas for a class project.

The chapters following concepts and background review delve into the details of photovoltaic and wind systems as interconnected or stand-alone designs. Estimating and predicting energy production is presented using industry distribution functions and online programs. Concepts of temperature coefficients, synchronization, power conversion, and system protection are explained and practiced. These concepts are applied to residential and small commercial systems and later extrapolated to large system design.

Economic analysis is presented using basic methodologies such as payback and rate of return. The methodology to develop advanced analysis is introduced using spreadsheets. A course on engineering economic analysis is recommended for students to develop a sound understanding of investment risk, rate of return, and return on investment.

The remainder of the text explores other renewable technologies, energy storage systems, thermal systems, and renewable related topics. This book is intended as a "hands-on" guide and is structured to motivate the student to experience the design and installation process. While students develop their personal design ideas, the course integrates field trips and guest speakers to enhance the educational experience.

KEYWORDS

energy, energy storage, power, photovoltaic, renewable, solar, thermal energy, wind energy

CONTENTS

CHAPTER 1

CONVENTIONAL ELECTRIC POWER SYSTEMS

Power and energy are important concepts that must be understood before beginning to explore the application of renewable energy (RE) systems. For example, a common misunderstanding is the confusion between "watts" (W) and "watt-hours" (Wh). Watts is a power term, and watt-hours is an energy term. Adding to the confusion are shortcuts used by power industry experts who frequently use the power term "megawatt" when they really mean "megawatt-hour" (energy). Likewise, a utility energy trader in dispatch will almost universally use the term "megawatt" when buying or selling a "megawatt-hour" of energy. We begin this book with definitions and the relationship between energy and power, as used by electrical power engineers.

1.1 POWER ENGINEERING CONCEPTS AND TERMINOLOGY

Power is defined as the time rate at which work is done or energy is emitted or transferred. *Energy* is the capacity for doing work. The *watt* is a unit of power and 1 W is equal to 1 joule per second—or in electric terms, 1 ampere under the pressure of 1 V. (*Joule* is the unit of mechanical power equal to a meter-kilogram-second or 0.7375 foot pounds.) The important fact to remember from these definitions is that power is a rate and energy is a quantity. Therefore, a *watt* is the rate at which energy is being produced or transferred. *Watt-hour* is the quantity of energy transferred or produced.

Electrical power equipment is typically rated in terms of watts. A 100 W light bulb will use 100 Wh of energy when operated for 1 h at

1 A and 100 V. A photovoltaic (PV) panel rated at 100 W will generate 100 Wh in 1 h when operated at a specified solar intensity and connected to an appropriate load. If these devices operate for 6 min, the resulting energy will be 10 Wh. A *kilowatt* (kW) is 1,000 W, a *megawatt* (MW) is 10^6 W, and a *gigawatt* (GW) is 10^9 W.

Power plant generators usually are operated at a constant power near the nameplate rating. This makes it simpler for dispatch operators to communicate by referring to 100 MW and assuming that the generator will operate at 100 MW for 1 h to produce a 100-MWh block of energy. We cannot make that assumption with PV systems or wind generators, because the output power is variable. Wind and solar power must be integrated over time to determine energy.

The *British thermal unit* (BTU) is a common measure of energy. One BTU is equal to the energy required to heat 1 pound (lb.) of water by 1 degree Fahrenheit (°F). A BTU is also equal to 0.293 Wh, and 3,412 BTUs equal 1 kWh. The BTU provides a useful conversion between electrical and solar thermal systems. The gas industry commonly uses the acronym MBTU to represent 1,000 BTUs or MMBTU to represent 1,000,000 BTUs. (The M is derived from the Roman numeral value.) This textbook will avoid using MBTU, as it causes frequent calculation errors and confusion. K represents 1,000 and M equals 1,000,000, as used in kW and MW.

As mentioned above, a joule is equal to 0.7375 foot-pounds (ft-lbs), and a joule per second is equal to 1 W. From these relationships, we can determine 1 kWh of energy to be equal to 2,655,000 ft-lbs. If a 150 lb. person climbs a 17,700-foot mountain, the amount of energy expended is about 1 kWh! Most of us use that amount of energy for home lighting every day! Another look at the value of 1 kWh is the content of energy in petroleum—approximately three ounces of oil contains 1 kWh of energy if we assume that 1 gallon of oil contains 143,300 BTUs. By comparison, if we install a 2 m² PV panel on our roof, it will generate about 1 kWh of electricity per day [1].

Demand is defined as the amount of work to perform a desired function. In the electrical world, the term "demand" is used to measure the peak power required to operate all connected loads for a particular circuit. If a circuit consists of five 100 W light bulbs, the peak demand is 500 W. Demand is determined when the load is at maximum, or when all five lights are on. Since utilities are concerned with the maximum load on their system, they measure peak energy in a specified time interval for demand. An instantaneous value is not as significant as the average value during this time interval—usually 15 min. For example, if the power on a circuit ranges from 5 to 15 kW during the 15-min interval, the demand will be about 10 kW. The average value of 10 kW is more significant than the 15 kW peak value, because it more accurately represents the load their generators must meet.

Capacity is the instantaneous ability to provide energy required to do work. Again, the electrical use of the term "capacity" usually refers to the size of the generator(s) required to maintain a circuit load. A 100-MW generator will provide the energy required to operate 100 MW of peak load. Capacity is also used to define the load capability of other electrical equipment on the system. A 115 kV transmission line may safely transfer 100 MW of power without overheating and sagging. A 100-megavolt-am-pere (MVA) transformer may safely transform 90 MW of power before it overheats. The safe operating rating of the equipment defines its capacity.

Using the electrical definitions of capacity and demand, *energy* can be defined as the product of demand and time-in-use ($D \times t$), or product of capacity and time-in-use ($C \times t$). These definitions are used when calculating total consumption. For example, the United States consumed about 4 trillion kWh of energy in 2005 [2]. That equates to about 13 MWh per year per person. The peak electrical demand in the United States is about 760 GW.

When scientists talk about global or national energy use, the term "quad" is frequently used. A *quad* is equal to one quadrillion BTUs. A quad is also approximately equal to the energy stored in 180 million barrels of crude oil. One quad equals 293 billion kWh. The total global primary energy production in 2004 was 446 quads.

Avoided costs are the incremental costs to an electric utility to generate electrical energy. These costs generally include fuel costs for generation, operational costs of transmission, and operational distribution costs per kWh. Fixed costs (e.g., capital equipment improvements, franchise fees) are not included in avoided costs.

Net metering generally refers to buying and selling energy at the same rate (Appendix D). Net metering applies to small (<10 kW) systems, where the financial impact is relatively negligible to the utility. Under net metering rules, an RE system is allowed to sell its generated energy at the same rate it buys energy from the utility. However, utilities and public utility commissions set rules for net metering, which vary by jurisdiction. The utility may require the RE entity to pay avoided costs for excess energy generated each month. More commonly, the utility will pay avoided costs for excess generation at the end of the year. Xcel Energy currently has an option to "bank" excess energy indefinitely. We will discuss the financial details of net metering in Chapter 7.

Renewable Energy Credit (REC) is an incentive for production of energy from RE systems. Investors can accumulate RECs from energy produced by their renewable systems. A political proposal has been suggested to impose carbon taxes on companies that generate carbon dioxide and other greenhouse gasses. RECs may be purchased to counter the carbon tax. Public utility commissions may also require utilities to include renewable generation in their mix of generation. The utility may then earn or purchase RECs to meet the portfolio requirements. The price on an REC varies widely, especially for residential PV systems. The REC contract price for small residential systems was about 10 cents/kWh in 2010 and 5 cents/kWh in 2014.

Production-based incentives (PBI) are becoming the standard for RE installation incentives. Instead of up-front cash incentives based on system size, utilities now offer incentives based on energy production over a fixed number of years. For example, in 2010 Xcel Energy offered an REC of 10 cents/kWh over the first 10 years of operation. The advantage of PBI to the utility is improved cash flow for the rebate program. Incentives are paid out during the same time period that the utility is collecting RE standard billing fees from the customers, which is called the renewable energy standard adjustment (RESA). PBI also ensures that the RE systems are operating as designed. The utility does not need to enforce REC production that is specified in operating contracts. Customers with RE systems will ensure that the system is operating as designed to keep those PBI incentive checks coming. A disadvantage of PBI is that an additional meter is necessary to measure all RE generation, and billing is more complex for customers with RE systems.

Production Tax Credit (PTC) for wind systems is a federal incentive that rewards investment in large wind generation. The current PTC rate is 2.2 cents/kWh. Since this is a federal incentive, it must be approved by Congress. Recently, the PTC incentive was extended for one year. As wind energy costs become more competitive with conventional power generation, the incentive will decrease and eventually dissolve.

Power Purchase Agreements (PPAs) are generally reserved for commercial- and utility-scale projects. Energy rates are determined in the negotiation for contract. The interconnecting utility will offer rates based on competitive sources of energy and the retail rates established within the jurisdiction by the regulatory agency. PPAs are used by independently owned utilities or publicly owned facilities.

Other acronyms are included in the Acronym and Abbreviations section at the end of this textbook. A glossary is also included with short definitions of terminology used in this textbook.

1.2   ELECTRIC POWER SYSTEM DESIGN

Conventional electric power systems in this country consist of generation, transmission, distribution, and loads. Generation is a mix of baseline, peaking, spinning reserve, offline reserve, and renewable (variable) equipment. Baseline refers to coal, nuclear, oil, and gas-fueled generators that run constantly except for scheduled maintenance. These are the largest generators that require relatively long start-up procedures to get them up to speed and online. Nuclear and coal facilities are usually baseline. Peaking generators are smaller and can be started and brought online in a relatively short time to assist with peak daily or seasonal loads. They are usually fueled with oil or natural gas. Spinning reserve units are large enough that it is more economical to keep them connected to the electrical grid and running at a low output and increase output to match load conditions. Any generator that is running at low output is not as economical as at full load and is avoided by utilities if possible. Offline reserve generators are similar to peaking generators but are usually larger and operated seasonally, or during scheduled maintenance times for baseline generators. Renewable systems, including wind and solar, are variable and must be supplemented with reserve or peaking generators when wind is low or solar energy is limited. For this reason, utilities try to limit the *penetration factor*, which is the ratio of grid-connected renewable power to conventional power.

Figure 1.1 is a one-line diagram of a power system including generators, transformers, transmission lines, and a load distribution center. G1 and G2 generators provide system power that is stepped up to transmission line voltage by transformer 1. T1 and T2 are transmission lines between two substation buses. Transformer 2 steps the voltage down for distribution and the load. Circuit breakers provide disconnect means and protection from system faults.

Transmission systems get electrical energy from the source (generation) to the load distribution centers. Transmission lines operate at much higher

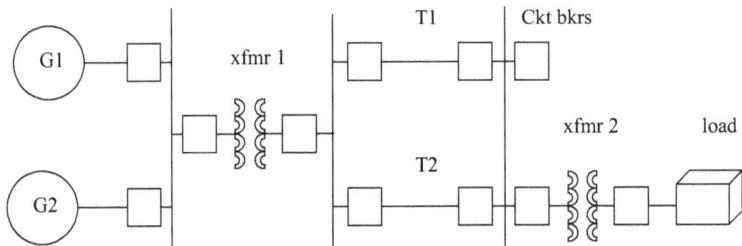

Figure 1.1. Conventional power system one-line diagram.

voltages than generation or loads to increase efficiency. A higher voltage means lower current for a given power. Lower current decreases power loss in the wire, and line capacity is proportional to the square of the voltage. With hundreds of miles of transmission lines, power loss is a significant factor in system operation. Typical alternating current (AC) transmission voltages are 230, 345, 500, and 765 kV [3]. A direct current (DC) voltage of 1,000 kV is used for long-distance transmission—350 miles or more. Transmission is a critical factor for many of our wind generation systems, because the wind farms are installed many miles from major cities and load centers. Furthermore, some large PV systems are being installed far from major load centers and require additional transmission line capacity.

Distribution load centers use large step-down transformers to lower transmission line voltage for power lines and equipment that distributes power to the customer. Distribution voltages range typically from 4 to 13.8 kV. Circuit breakers are installed to provide switching and disconnect means for load control and system protection. System protection equipment has become very sophisticated in the last few years. Digital protection equipment can detect faults within one cycle (1/60 s) and open circuits before system damage or instability occurs. The addition of variable RE systems to the grid increases the complexity of protection systems. Since most PV systems and wind generators have significantly less fault current capacity, the protection equipment must be more sensitive to voltage and power fluctuations. Electrical power engineers have been able to improve the operation of the electrical grid to accommodate renewable generation systems to date. However, the caveat exists that higher penetrations of renewable systems may cause disturbances the existing power grid cannot absorb, and widespread outages may occur.

The fuel source for most of our conventional electric power generation system is coal, which is the main contributor to carbon dioxide (CO_2) emission, pollution, and other externality costs. Coal is the fuel source for 41% of power generation in the United States, followed by natural gas (21.3%), hydro (16%), nuclear (13.5%), and oil (5.5%) [4]. In contrast, coal provides about 25% of the world's primary energy, and oil contributes about 34% [5]. These fossil fuels contribute to worldwide pollution and carbon dioxide generation. Externalities are the consequences of activities that are not normally part of the economic evaluation of electric generation. For example, the costs attributed to health problems caused by pollution are not directly paid by the generation provider (i.e., costs are external to the plant operator). Currently, new coal plants in the United States are required to include expensive pollution control equipment (e.g., scrubbers) that mitigates pollution. However, CO_2 production continues to increase as more coal plants are built and pollution is generated—and environmental concerns are inevitable.

From 2008 to 2010, 16 new coal plants were built in the United States. Another 16 are now completed or under construction. They will generate 17.9 GW—enough to provide power for about 15.6 million homes. They will also emit 125 million tons of greenhouse gases per year. The present federal administration has set aside $3.4 billion in stimulus spending for "clean coal" technology to capture and store greenhouse gases. None of the new plants is built to capture CO_2 emissions, although the U.S. Department of Energy has spent $687 million on clean coal programs and $35 billion has been invested in these traditional coal plants.

Another concern with our conventional fuel source is environmental damage caused by mining, drilling, and transportation. The Exxon tanker Valdez oil spill and the British Petroleum offshore drilling explosion and leak are highlights of externalities associated with the oil industry. The oil shale industry is also experiencing contamination problems due to hydraulic fracturing (i.e., fracking)—the injection of fluids into deep geological formations to recover oil and gas reserves. Hydraulic fracturing enables the production of natural gas and oil from rock formations deep below the earth's surface (generally 5,000–20,000 feet). At this depth, there may not be sufficient porosity and permeability to allow natural gas and oil to flow from the rock into the well bore and be recovered. A fluid is pumped into the rock to facilitate release of the trapped reserves. The composition of the fluid is proprietary information that is not made readily available to the public. The fluid used in fracking may be spilled or work its way into water supplies. Although drilling companies report that the fluid and process are safe, there are many contrary claims. Current studies are underway to determine if environmentally dangerous gasses are also released in the fracking process.

1.3   ELECTRIC POWER ANALYSIS

Before we can explore RE power systems, we must review voltage and current representation for conventional power systems operating under normal conditions. The conventional AC power system consists of purely sinusoidal voltage and current waveforms when in a steady state condition. This chapter reviews typical representations of voltage, current, and power values in single-phase and three-phase configurations.

The equations derived in this chapter will be used throughout the book to analyze RE systems interconnected to conventional power systems. Power factor, real power, reactive power, and apparent power are particularly useful components that are used to analyze any power system. Per-unit analysis, voltage drop calculations, distortion measurement, and power flow analysis are additional tools introduced in this chapter, which are useful in succeeding chapters to analyze power system integration.

1.3.1 SINGLE PHASE CIRCUITS

AC voltage for a single-phase circuit is represented by *a sinusoidal time function v(t)*, as shown in Figure 1.2.

$$v(t) = V_{max} \cos(\omega t + \theta_r) \tag{1.1}$$

$$v(t) = 141.4 \cos(377t + 0) \tag{1.2}$$

where V_{max} is the peak value of the waveform (141.4 V). ω represents *angular velocity* and is equal to 377 ($\omega = 2 \times \pi \times 60$) for a 60-cycle per second (Hz) system. θ_r indicates the *phase angle* between an instantaneous value and a reference time (e.g., $t = 0$). The phase angle for the voltage waveform shown is zero. θ_r must be expressed in radians since ωt is in radians (radians = degrees $\times \pi/180$).

A more convenient representation of voltages, currents, and power can be shown in a *phasor diagram* that displays magnitude and the instantaneous angle from a reference (usually voltage at 0° or 0 radians). Figure 1.3 shows the same voltage as in Figure 1.2, but with a 30° phase angle. The common electrical convention for phase angle is 0° on the positive *x*-axis and 180° on the negative *x*-axis.

$$V = |V| \angle\theta = 100 \angle 30° \tag{1.3}$$

v(t) = Vm cos (wt)

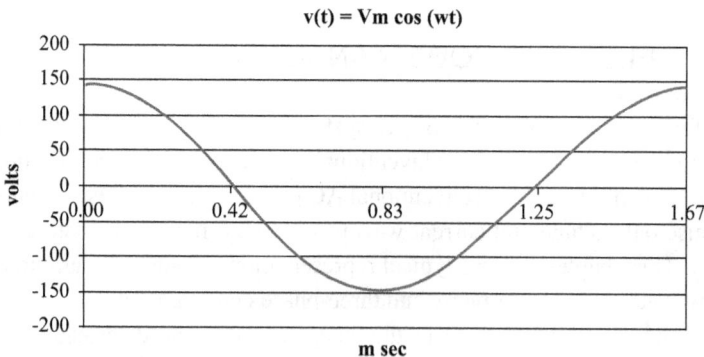

Figure 1.2. AC voltage waveform.

Phasor representation of the voltage waveform is given as a magnitude and phase angle. Phasor voltages are the root-mean-square (rms) values of voltage or $V_{max}/\sqrt{2}$. The bars enclosing the phasor ($|V|$) indicate rms values of the voltage or current. RMS values are the effective values of voltages, currents, and power. The *effective value* of power is the average power expended in a resistor ($P = |I^2| R$). These are the values read by ordinary voltmeters and ammeters.

Another common representation for voltage, current, and impedance is the *rectangular* format (Figure 1.4):

$$V = A + j\,B \quad V = 86.6 + j50 \text{ V,} \tag{1.4}$$

where $A = |V| \times \cos(\theta)$ and $B = |V| \times \sin(\theta)$.

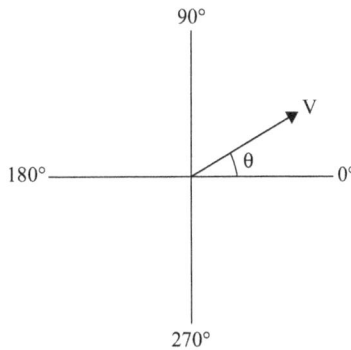

Figure 1.3. Phasor representation of voltage.

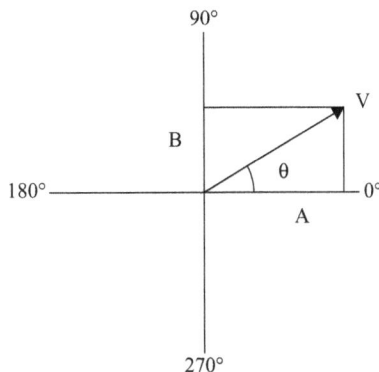

Figure 1.4. Rectangular representation of voltage.

The rectangular format may also be used to represent impedance (Figure 1.5):

$$Z = |Z|\angle 30° = |100|\angle 30°$$

$$Z = R + jX \text{ ohms} = 86.6 + j50 \text{ ohms}, \tag{1.5}$$

where $R = |Z| \times \cos(\theta)$ and $X = |Z| \times \sin(\theta)$.

Now, let us look at the current waveform, which is similar to the voltage sinusoidal waveform. The maximum peak value of the current waveform is 14.4 A and the phase angle in Figure 1.6 is zero (maximum value at $\omega t = 0$). A cosine function is used in this example.

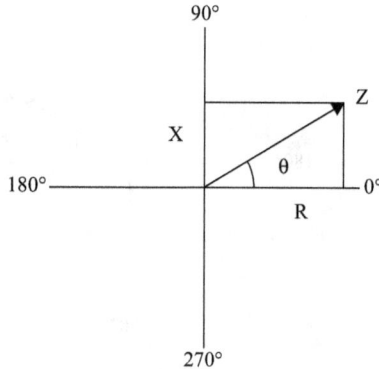

Figure 1.5. Rectangular representation of impedance.

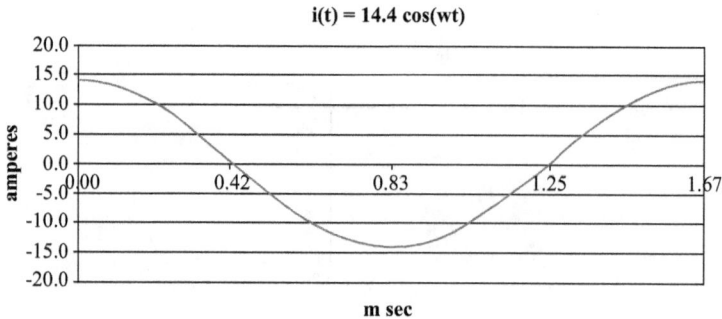

Figure 1.6. Time representation of current.

If we combine voltage and current time function waveforms and add a lagging phase angle of 30° to the current, the result is given in Figure 1.7. Note that the zero crossing of the current waveform is 0.14 ms after the voltage zero crossing, which corresponds to 30° lagging current.

Now we can talk about power and power factor. Power is equal to the product of voltage and current. The power waveform, $p(t) = v(t) \times i(t)$, is shown in Figure 1.8. The 30° lagging current creates a power waveform that is not in phase with the voltage. Instantaneous power is negative when the polarity of current and voltage is opposite.

If voltage and current are in phase (i.e., $\theta = 0$), the power waveform is in sync with the voltage and current, but twice the frequency. This means that the load is purely resistive (i.e., volt-amps = watts). It can be easily shown that if the voltage and current are 90° out of phase, the resulting power waveform would average out to be zero, and no watts are used by the load (i.e., volt-amps = reactive volt-amps). A lagging current is caused

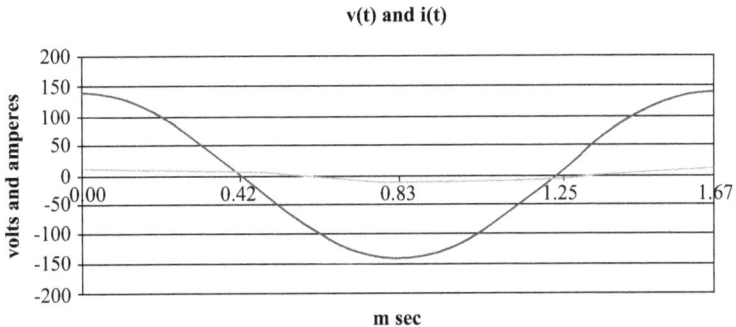

v(t) and i(t)

Figure 1.7. Time representation of voltage and lagging current.

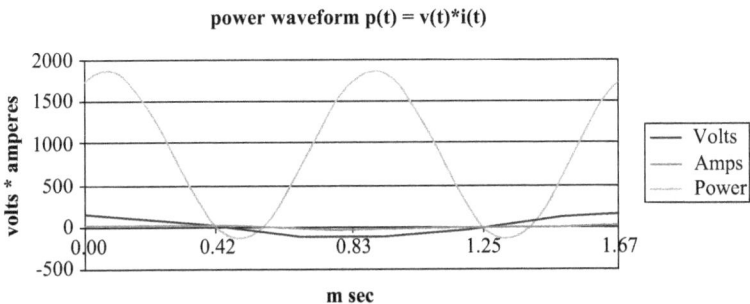

power waveform p(t) = v(t)*i(t)

Figure 1.8. Time representation of power.

by an inductive component of load and creates a lagging power factor. Similarly, a leading current is caused by a capacitive component of load and creates a leading power factor. Leading or lagging is relative to the applied voltage.

1.3.2 POWER CALCULATIONS

Power factor (PF) is calculated from the cosine of the angle (θ) between voltage and current (PF = cos(θ)). For our 30° phase angle, PF = cos(30) = 0.866. PF is defined as the resistive component of power divided by the total apparent power. *Apparent power* (S) is the product of rms voltage and current ($|V| \times |I|$). *Real power* (P) is the product of rms voltage and current times the power factor or cosine of the phase angle, that is, $P = |V| \times |I| \times \cos(\theta)$ or $P = |V| \times |I| \times$ PF. *Reactive power* (Q) is the component of apparent power that flows alternately toward or away from the load, that is, $Q = |V| \times |I| \times \sin(\theta)$. Keeping in mind that S, P, and Q are rms values, we may simplify the power equations as follows:

$$S = V \times I \tag{1.6}$$

$$P = V \times I \cos \theta \tag{1.7}$$

$$Q = V \times I \sin \theta \tag{1.8}$$

The derivations of P, Q, and S using trigonometry can be found in basic textbooks on power [6]. Cosine laws are used to convert Equation 1.9 to Equation 1.10, which shows the real and reactive components of the apparent power (p).

$$p(t) = V_m \cos wt \times I_m \cos(\omega t - \theta) \tag{1.9}$$

$$p = [V_m \times I_m/2] \cos \theta (1 + \cos 2\omega t) + \tag{1.10}$$
$$[V_m \times I_m/2] \sin \theta \sin 2\omega t,$$

where the real component is $[V_m \times I_m/2] \cos \theta (1 + \cos 2\omega t)$ and the reactive component is $[V_m \times I_m/2] \sin \theta \sin 2\omega t$.

Real and reactive power can be shown with much more clarity using a power triangle. Figure 1.9 shows a power triangle and the relationships of P, Q, and S:

$$S^2 = P^2 + Q^2 \tag{1.11}$$

$$P = [S^2 - Q^2]^{1/2} \tag{1.12}$$

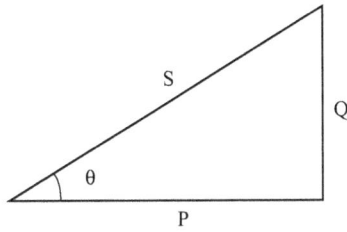

Figure 1.9. The power triangle.

Power factor can be obtained from P and S: PF $= P/S$. Trigonometric functions can also be used to determine the power factor:

$$PF = P/[P^2 + Q^2]^{1/2} \qquad (1.13)$$

$$PF = \cos[\tan^{-1}(Q/P)] \qquad (1.14)$$

If the phase angle is zero, the load is purely resistive and $P = S$. Similarly, if the phase angle is 90° or 270°, the load is purely inductive or capacitive.

By convention, a positive value of Q is an inductive load and a negative value of Q is a capacitive load [3]. The general engineering conception is that a capacitor generates positive reactive power, which supplies Q required by an inductive load. However, it is more convenient to consider that the reactive power is positive when supplying an inductive load. The concepts of power factor, reactive power, and real power will become significant in future discussions on RE sources and interconnection with conventional power systems. Reactive power does not require "coal off the coal pile" since reactive power flows back and forth but reduces the capacity of the system and increases transmission losses. The effects of power factor and reactive loads will be discussed later in relation to inductive wind generators and interconnected solar systems.

Another useful tool for calculating S, P, and Q is *complex power equations*. If the magnitude and phase angles for voltage (α) and current (β) are known, apparent power may be calculated using the complex conjugate:

$$S = P + jQ = VI^* \qquad (1.15)$$

In this equation, I^* is simply the magnitude of I and its conjugate (magnitude with negative phase angle). Apparent power is the product of the magnitudes of V and I and the sum of the voltage phase angle and the negative of the current phase angle.

$$S = V \times I \angle (\alpha - \beta) \qquad (1.16)$$

1.3.3 PER-UNIT ANALYSIS

Another tool that simplifies power calculations is per-unit analysis. Instead of using the actual voltage in calculations, a base voltage and a base apparent power are identified and set equal to 1.0 (PU). Then all calculations may be expressed on the base reference. For example, if a base voltage of 120 kV is chosen, voltages of 108, 120, and 126 kV become 0.90, 1.00, and 1.05 (PU). We may also express these values as 90%, 100%, and 105%. However, percent becomes cumbersome when multiplying two quantities. The product of 1.2 PU voltage and 1.2 PU current becomes 1.44 PU volt-amperes. The product of 120% voltage and 120% current is 14,400, and we must divide this by 100 to get the proper answer of 144%.

Voltage, current, kilovolt-amperes (kVA), and impedance are so related that selection of a base value for any two of them determines the base values of the remaining two. For example, if we select 12 kV as our base voltage, and 100 kVA as our base apparent power, our base current may be determined by dividing apparent power by voltage:

Base I = base kVA/base kV = 100 kVA/12 kV = 8.3 A $\qquad (1.17)$

Per-unit calculation simplifies number crunching and allows relative comparisons of values of systems with varying voltages and power. One must be careful to separate single-phase systems from three-phase systems by identifying whether the base is a single-phase quantity or the sum of all three phases.

1.3.4 EXAMPLE PROBLEMS

Example Problem 1.1

A commercial load is connected to the grid through a demand meter. The meter shows a load of 1,000 kW and 450 kvar of reactive power. What is the power factor?

Since we are given real (P) and reactive power (Q), we may use equation 1.13 to determine the power factor angle. The arctangent of Q/P is 24.2°, and the cosine of 24.2° is 0.912, which is the power factor.

Example Problem 1.2

What is the apparent power (S) of the load in Example Problem 1.1?

Solution:

Applying Equation 1.11, S is equal to the square root of the sum of the squares of P and Q, or 1,096.6 kVA.

Example Problem 1.3

If the commercial load is offset with a PV system that generates 500 kW of real power (power factor of 1.0), what is the new power factor at the meter?

Solution:

Recalculating the power factor for $P = 500$ kW and $Q = 450$ kvar, PF = 0.74

Example Problem 1.4

If the utility requires a minimum power factor of 0.85, how many kvars of capacitive correction must be added to the commercial system in Example Problem 1.3?

Solution:

Assuming that the real power remains at 500 kW, and the power factor is now 0.85, apparent power is $S = P/\cos\theta$ or $S = P/PF = 588$ kVA. $\theta = $ arc cos $(0.85) = 31.8°$. Reactive power (from the power triangle) $Q = S\sin\theta = 588 (0.527) = 310$ kvar. The capacitive bank must add 140 kvar (450–310) to the commercial system bus.

1.4 THREE-PHASE POWER SYSTEMS

Section 1.3 analyzed single-phase power systems using basic mathematical and trigonometric methods. Now we will apply those methods to a three-phase system. Utility generators, transmission, and distribution systems consist of three-phase equipment. In addition, most commercial and industrial equipment is three phase. Three-phase motors and generators are more practical and efficient than single-phase machines. Utility generators are designed to produce balanced three-phase voltages that are 120° out of phase. The three windings of a generator are constructed with 120° mechanical orientation of the windings. Figure 1.10 shows a basic three-phase generator and load for a balanced system:

The generator is wound for a wye connection with a common neutral o. The load is also wye connected with a neutral connection n. Z_g is the generator internal impedance, which is inductive and with low resistance. Z_l is the load impedance, which may consist of resistance, inductance, and capacitance. Figure 1.10 shows an inductive load with only resistance and inductance.

We will assume a *balanced* system to simplify calculations and simulate ideal operating conditions. This means that the magnitude of generator voltages are equal $|V_a| = |V_b| = |V_c|$ and 120° in phase difference: $V_a = V_b \angle 120° = V_c \angle 240°$. Current is also equal in magnitude with a 120° phase shift, and the sum of the three-phase currents is zero ($I_n = I_a + I_b + I_c = 0$). Any imbalance in the phase loads will create a neutral current, which is unwanted and complicates system analysis.

Time-function voltage waveforms for phases a, b, and c are 120° out of phase as shown in Figure 1.11. The sum of the three voltages at any instant in time will equal zero ($V_n = V_a + V_b + V_c = 0$). The current waveforms are

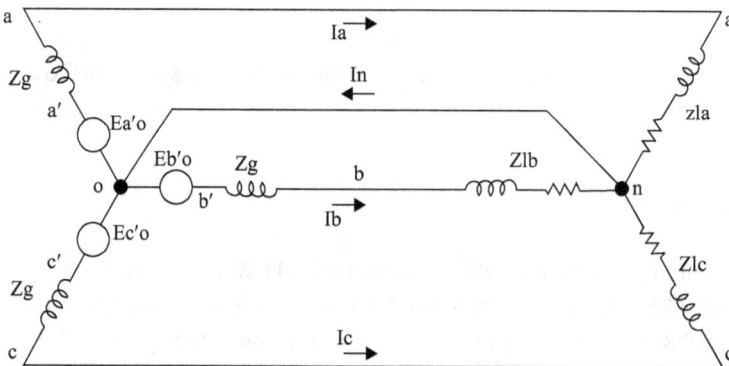

Figure 1.10. Three-phase system diagram.

also 120° out of phase for a balanced system and will sum up to zero at the neutral point.

The phasor diagrams (Figures 1.12 and 1.13) represent the phase relationship of the three-phase system more clearly. Conventional rotation of the generator will create a phase sequence of abc.

The phase relationships of V_a and I_a are shown with a 90° phase difference. This would indicate that the load is purely inductive. Normally, the phase current (I_a) would lag the phase voltage by a few degrees with a primarily resistive load with small inductance.

We may also decide to use a line-to-line voltage (V_{ab}) for the reference voltage instead of the line-to-neutral voltage (e.g., V_a) as shown above. As a phasor, V_{ab} leads V_a by 30°. Figure 1.14 shows the geometry for V_{ab} and

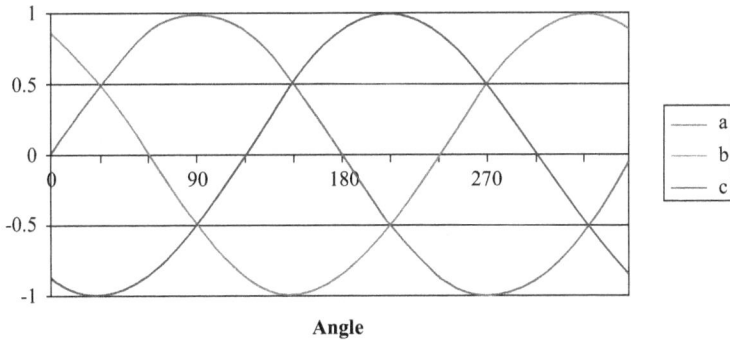

Figure 1.11. Three-phase waveforms.

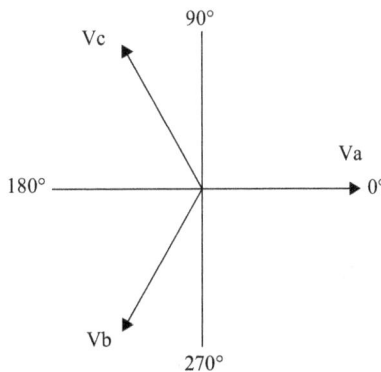

Figure 1.12. Voltage phasor diagram.

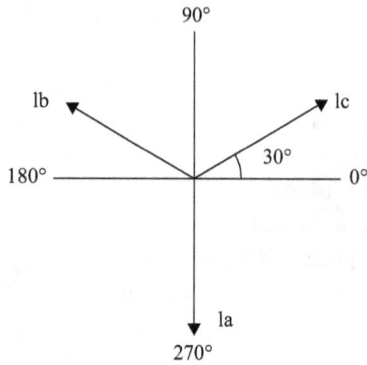

Figure 1.13. Current phasor diagram.

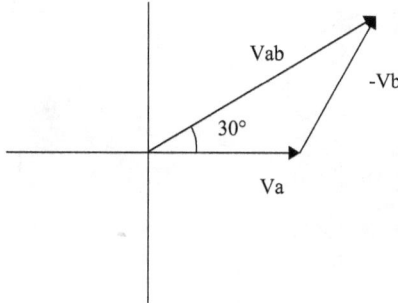

Figure 1.14. Line-to-line voltage.

the line-to-neutral voltages V_a and V_b. The magnitude of V_{ab} is calculated using Equation 1.18:

$$|V_{ab}| = 2|V_a| \cos 30° = \sqrt{3} \, |V_a| \tag{1.18}$$

Figure 1.15 shows the line-to-line voltages (V_L) for a balanced three-phase system. Line-to-line voltage is commonly used as a reference for all three-phase systems. For example, a 115 kV transmission line and a 480 V commercial distribution system have line-to-neutral voltages of 66 kV and 277 V, respectively. The magnitude of line voltage is equal to the product of phase voltage and the square root of 3. Phase current is the conventional reference for line current for wye-connected systems.

$$V_L = \sqrt{3} \, V_\varphi \tag{1.19}$$

$$I_L = I_\varphi \tag{1.20}$$

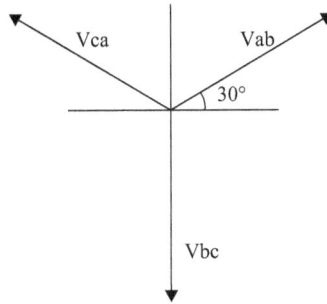

Figure 1.15. Three-phase line-to-
line voltages.

1.4.1 POWER IN BALANCED THREE-PHASE SYSTEMS

With the conventions used in Section 1.4, we calculate *three-phase power* using the line references [7].

$$P = \sqrt{3}\ V_L I_L \cos \theta_p \tag{1.21}$$

The power phase angle (θ_p) is the angle by which the phase current lags the phase voltage. *Apparent power* (*S*) is total volt-amperes, and *reactive power* (*Q*) is the inductive or capacitive component of apparent power. They are calculated using Equations 1.22 and 1.23:

$$S = \sqrt{3}\ V_L I_L \tag{1.22}$$

$$Q = \sqrt{3}\ V_L I_L \sin \theta_p \tag{1.23}$$

Equations 1.21, 1.22, and 1.23 are commonly used to calculate *P*, *Q*, and *S*, respectively. Unless identified otherwise, voltages are assumed to be line-to-line, currents are line, and power is for all three phases. Again, balanced conditions are assumed unless specified otherwise.

A *power triangle* (Figure 1.16) is very useful to show the trigonometric relationship of apparent power, real power, and reactive power. The relationships are identical to those given in Section 1.3 for single-phase circuits.

$$S^2 = P^2 + Q^2 \tag{1.24}$$

$$P = [S^2 - Q^2]^{1/2} \tag{1.25}$$

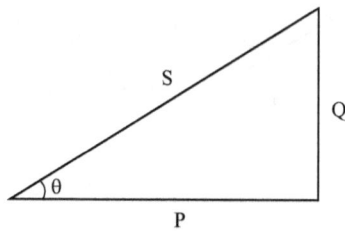

Figure 1.16. Power triangle.

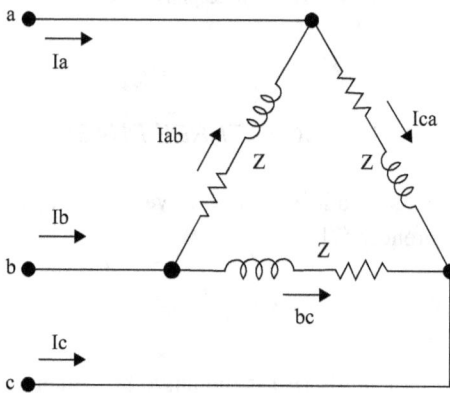

Figure 1.17. Delta connection.

1.4.2 WYE AND DELTA CONNECTIONS

For simplicity, the above discussion used wye connections at the generator and load. Delta connections are common in three-phase generators and transformers. If the load shown in Figure 1.10 is replaced by a delta connected load, the load is represented by Figure 1.17. Although the currents in the delta windings are not the same as the line currents, the power calculations do not change, if care is taken to correctly identify line current (I_a) and line voltage (V_{ab}). Also, note that a neutral connection is not available with delta windings. Therefore, neutral current cannot flow between generation and load.

When solving three-phase *balanced* circuits, it is not necessary to work with the entire three-phase circuit as shown in Figure 1.10 or 1.17. The circuit may be solved using Kirchhoff's voltage law around a closed path that includes one phase and neutral. If the load is connected in delta, it is easy to convert the delta connection to its equivalent wye and proceed

Equivalent circuit

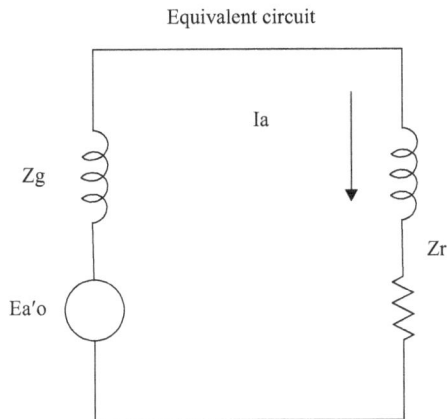

Figure 1.18. Equivalent circuit.

with the calculation of a single phase. The impedance of each phase of the equivalent wye will be one-third the impedance of the delta that it replaces. The equivalent circuit is shown in Figure 1.18 for phase "a" of Figure 1.10:

Voltage, current, and power calculations may now be easily done using the equivalent circuit. When converting back to three-phase power, add the power in each of the three (balanced and equal) phases. Remember to include phase relationship and power angle in the calculation. Total three-phase power is given in Equation 1.21.

1.4.3 UNBALANCED SYSTEMS

Unbalanced systems require analysis by more sophisticated tools, since the trigonometric functions we have discussed become complex in an unbalanced three-phase system. In 1918, a powerful tool called *symmetrical components* was developed by Fortescue [8]. According to Fortescue's theorem, three unbalanced phasors can be resolved into three balanced systems of phasors. The positive sequence of components consists of three phasors equal in magnitude and displaced by 120°, with the same sequence as the original phasors. The negative sequence differs in that the sequence is opposite, and the zero-sequence phasors have zero displacement from each other. A fault or disturbance on the system can be analyzed using balanced system techniques for each sequence. Symmetrical component analysis is used universally by modern system protection equipment to identify faults and control switching equipment. Further description of this technique is beyond the scope of this text, and balanced systems will be assumed.

PROBLEMS

Problem 1.1

A 100 MW coal-fired power plant generator runs continuously at rated output (100 MW) for 30 days. How much energy (kWh) is produced in the 30 days?

Problem 1.2

If a home uses 20,000 BTUs per day for heating, what is the equivalent energy used in kWh?

Problem 1.3

A 10 kW PV system is expected to generate about 15,000 kWh/y. If the customer has a net metering agreement with the utility and uses all the energy generated, what are the savings if energy is valued at 10 cents/kWh? If this customer also receives a PBI of 9 cents/kWh, what is the PBI revenue per year?

Problem 1.4

What is the generation mix for the utility serving you?

Problem 1.5

A single-phase voltage of 240 V is applied to a series circuit with an impedance of 10 $\angle 60°$ ohms. Find R, X, P, Q, and the power factor of the circuit.

Problem 1.6

If a capacitor is connected in parallel with the circuit of Problem 1.5, and the capacitor supplies 1,250 vars, find the new P and Q supplied by the 240 V source. What is the new power factor?

Problem 1.7

In a balanced three-phase system, the line-to-line voltage is 480 V. Y-connected impedance is $10\angle30°$ per phase. What is the line-to-neutral voltage (V_φ) and phase current (I_φ)? Express the current in rectangular and polar format.

Problem 1.8

Determine the current in a three-phase motor when voltage is 480 V and the motor is running at full rated output of 15 HP. Assume the efficiency is 100% and power factor is unity.

Problem 1.9

If the power factor is 80% lagging, calculate P and Q for the voltage and current given in Problem 1.8.

Problem 1.10

In a balanced three-phase distribution line, the line-to-line voltage is 13.8 kV. The line is Y-connected at source and load. What is the phase-to-neutral voltage (V_φ)?

Problem 1.11

The distribution line of Problem 1.10 is 15 miles long. The impedance of the line per phase is $0.2 + j0.5$ ohms/mile. A load of $50 + j20$ ohms is connected in Y. What is the phase current (I_φ)? What is the total apparent, reactive, and real power provided by the source (generator)?

Problem 1.12

Determine the power factor at the generation source in Problem 1.11. Calculate the voltage drop from generator to load.

Problem 1.13

a. A utility power feeder supplies a customer with a load of 1,000 kW and 450 kvar. Using the power triangle and three-phase power equations, solve for the apparent power. What is the power factor at the customer's load?

b. The customer installs a 500 kW PV system at the utility point of interconnection. What is the new power and apparent power when the PV array is operating at full output (500 kW)? What is the new power factor?

c. The utility points out a lower power factor at the point of interconnection and claims that the customer is causing system inefficiency because of the lower power factor caused by connecting a PV system to the utility power system. Is the claim valid?

Hint: Using per-unit analysis, calculate voltage drop on a feeder that connects the PV array and load to the utility. Use the following assumptions:

$$PU \text{ (apparent power)} = 1,000 \text{ kVA}$$

$$R \text{ (line resistance)} = 0.05 \text{ PU } (5\%)$$

$$X \text{ (line reactance)} = 0.142 \text{ PU } (14.2\%)$$

Calculate the line voltage drop with and without the PV array connected. Do your results support the claim by the utility that a PV connection is detrimental to the utility system?

CHAPTER 2

Types of Renewable Energy Systems

Wind and photovoltaic (PV) systems are the most recognizable types of renewable energy systems we see today. There are many more sources of energy that are considered renewable including hydroelectric, solar thermal, marine, geothermal, and bioenergy. Many of these systems have been implemented for many years, such as hydroelectric systems. Hydro plants continue to supply large amounts of power to our electrical grid, and there is a potential for limited growth. Other systems are still in infancy although they have a huge potential, such as marine energy, which includes tidal power and wave power. This textbook will briefly describe some of the potential and limitations of the "other" renewable sources and provide references to some very good textbooks and documents that describe these technologies in detail.

Hydroelectric power plants currently supply about 20% of the world's electric generation [9]. Hydro generation is a very essential part of the electrical grid, because output can be controlled to compliment variable generation sources such as wind. Hydro plants are efficient at low generation levels and can be used to follow predicted and unpredicted changes in consumer demand. Hydro generators can respond within minutes to a change in demand. Pumped storage systems are also used to meet demand changes. Water is pumped to an upper storage reservoir at night when demand is low and released for generation during high demand times. Although many of the most attractive sites have been tapped with hydroelectric systems, there are locations where this technology may be developed (Figure 2.1).

Marine systems are composed of tidal and wave power generators. Tidal power has the potential to provide a large portion of world energy use, but efficient systems have not been developed. A problem with tidal

Figure 2.1. Hydroelectric facility. (*Source*: Photo courtesy of National Renewable Energy Laboratories (NREL).)

power is variability. Output generation depends on tidal frequency, which is approximately twice per day. Spring (in-phase relationship of the moon and sun) and neap (out-of-phase relationship) tides also cause further variability in power. Low head turbines and two-way generation are features that have not been developed for tidal power. Wave power systems also have a large potential to generate energy, but the technology and capital needed to capture the energy from oceanic waves have not developed. Some of the problems include the low frequency of large waves and the variability of wind by season. Scientists have projected that wave energy in our oceans has the potential of 2000 TW, or about twice the present world capacity.

Bioenergy or biomass systems are similar to fossil fuel systems in construction and operation. A boiler and steam turbine produce power from energy supplied by stored fuel with limited energy density. Fuel density limits the distance bio power plants can be located from the source, and the size of the plants is generally limited to 100–500 kW. Although direct combustion biomass systems may decrease the demand for imported oil, the process has several disadvantages. The fuel competes with food crops and may cause shortage or increase food prices. The environmental impact of CO_2 production and pollution also exists.

Solar thermal electric systems are in use at several locations in the United States. A large-scale plant near Barstow, California, has been in operation for about 20. The system uses mirrors to concentrate direct solar radiation for a boiler (see Figure 2.2), which drives the electric generator.

Figure 2.2. Solar thermal plant. (*Source*: Photo courtesy of NREL.)

Since the system requires direct radiation, the southwest is the preferred location for similar plants. Cloudy skies and diffuse radiation are unsuited for this technology. Parabolic trough collectors are also used to force radiation on a piped system that transfers heat to a heat exchanger/boiler where steam is produced to drive a generator. Xcel Energy is currently adding a solar thermal system to one of its coal-fired plants to supplement plant auxiliary power requirements. Smaller systems use parabolic concentrators that work at high temperatures and are suitable for distributed electric generation.

Wind power is the fastest growing energy source in the United States. In 2008, the U.S. wind industry installed 8,358 MW of new generation and took the lead in global installed wind energy capacity with a total of 25,170 MW [10]. In 2009, the United States added another 10 GW of wind capacity to reach a total of 35 GW. In 2012, 13.1 GW of wind energy was installed, which was more than natural gas generation. The U.S. Energy Information Administration estimates that U.S. electric consumption will grow to 3,902 billion kWh by 2030. If the United States is to meet 20% of that demand, the U.S. wind generation capacity will have to be greater than 300 GW. Much of this growth is attributed to the Renewable Energy Portfolio Standard (RPS) and the renewable energy production act (production tax credit [PTC]) bill passed by Congress in 2002. Utilities are required to produce 10% of their energy from renewable sources by 2010 [11]. The PTC provided a 1.8 cent/kWh incentive, indexed to inflation, for

Figure 2.3. Wind turbine generator at National Wind Test
Center. (*Source*: Courtesy of NREL.)

the first 10 y of wind plant production. The PTC is now at 2.2 cents/kWh
(Figure 2.3).

Geothermal systems use extreme hot water from deep wells for heating purposes or to generate electricity when the pressure is high enough
to drive a turbine. These systems are limited to areas where geothermal
sources can be reached economically with drilling equipment. Many
areas in the Rocky Mountains have large untapped geothermal resources,
but temperatures over 200°F are several thousand feet below the surface
(Figure 2.4).

Ground source heat pumps (GSHP) are sometimes confused with
geothermal systems. GSHP systems can be efficiently used for residential
heating and cooling. A network of piping 10–20 ft underground transfers
heat from the ground in the winter when earth temperatures are relatively

Figure 2.4. Geothermal plant. (*Source*: Courtesy of NREL.)

warm, and transfers heat from the home to the cool earth in the summer. The piping may also consist of 200 ft of vertical supply and return piping in the ground. Electric power is needed to run pumps and heat exchangers, which offsets some of the savings. A backup heating system may also be required during periods of very cold temperatures. Some utilities will subsidize GSHP systems to replace natural gas systems with electrical use. A solar electric system can supplement the electrical use with net metering or a storage (battery) system.

Solar electric systems using PV modules are also growing rapidly as the cost for photovoltaic (PV) modules drops, and residential and small commercial systems are subsidized by state government programs. A good measure of the growth of PV installations is the volume of PV shipments in the world. In 1980, less than 10 MW were shipped, and in 2006, more than 1,000 MW of PV modules were shipped [12]. In 2007, about 93% of PV modules were manufactured in Europe and Japan, and now China is a leader in PV manufacturing. The United States installed 3.2 GW of PV in 2012 (Figure 2.5).

The U.S. Energy Information Administration (EIA) provides yearly reports concerning energy generation. The latest report shows a 12.8% increase in nonhydro renewable generation in 2012 as compared to 2011. Now, nonhydro renewable generation is 5.4% of all U.S. electric generation sources. Coal fuel generation has decreased 12.5% to 37.4% of the

Figure 2.5. Residential PV system.

total mix. Nuclear energy is below 19% of U.S. electric generation. Natural gas generation has increased to 30.3% of electric generation. Solar and wind power is still a small portion of overall electric generation (0.11% and 3.46%, respectively).

CHAPTER 3

THE SOLAR RESOURCE

The sun generates enough energy in 1 h to meet all the needs on Earth in 1 y! If we were able to capture and store a small fraction of the sun's energy, we would not be dependent on fossil fuels. As sunlight enters the Earth's atmosphere, some is absorbed, some is scattered, and some passes through until it is absorbed or reflected by objects on the ground. The amount of radiation that is scattered or reflected depends on several factors, including the solar spectrum or wavelength of light, atmospheric conditions, and the surface on the ground.

This chapter reviews how we measure solar radiation available to us and how we design solar systems to maximize the capture of that solar energy. There are many factors that determine how much useful solar energy reaches our space on the earth's surface, including atmospheric conditions, the solar system geometry, shading, and array orientation. We will also introduce useful tools for the design of a solar system.

3.1   THE GEOMETRY OF SOLAR RADIATION

Sunlight, as we recognize it on Earth, consists of three components: direct, diffuse, and albedo radiation. *Direct radiation* reaches the Earth without scattering by the atmosphere. Direct radiation consists of direct rays and is sometimes called a light beam. It can be focused, as demonstrated by a magnifying glass. *Diffuse radiation* is the scattered component that reaches the Earth. Atmospheric particles and water vapor (clouds) create diffuse radiation, which is a significant component of total radiation, as verified on a cloudy day. *Albedo sunlight* is reflected from the ground or objects. The sum of all three components is called *global radiation*. At the top of Earth's atmosphere, global radiation is 1,367 W/m². At sea level, total global radiation is about 1,000 W/m², which means that the atmosphere scatters about 70% of the sun's radiation.

Zenith angle (ζ) is the angle between a direct beam from the sun and a line perpendicular to the Earth's surface (see Figure 3.1).The solar path for direct radiation is at a ζ of zero. A ζ of 90° results in zero direct radiation on the (parallel) surface.

Air mass index (AMI) is the relative path of direct solar radiation through the atmosphere. AMI affects the amount of spectral content of solar radiation reaching the Earth's surface and varies with sun's position and altitude. A direct path ($\zeta = 0$) correlates to an AMI of 1 (i.e., AM 1). The AMI above the atmosphere is zero (AM 0) and represents extraterrestrial radiation. AM 1.5 represents the solar path when the sun is at a solar altitude of about 42°, which is equal to a ζ of 48.2°. Figure 3.1 shows an AMI of 1.5.

The AMI is proportional to the secant (1/cos ζ) of the ζ and dependent on the composition of the atmosphere. Different air molecules absorb or reflect different sunlight wavelengths. The AMI is a general reference used to standardize global radiation measurements. AM 1.5 is generally used to rate a photovoltaic (PV) module or cell. The AMI may also be corrected for altitude, given atmospheric pressure at the proposed site. The following formula is used to calculate the AMI:

$$AMI = [1/\cos \zeta] \times [P_a/P_o] \qquad (3.1)$$

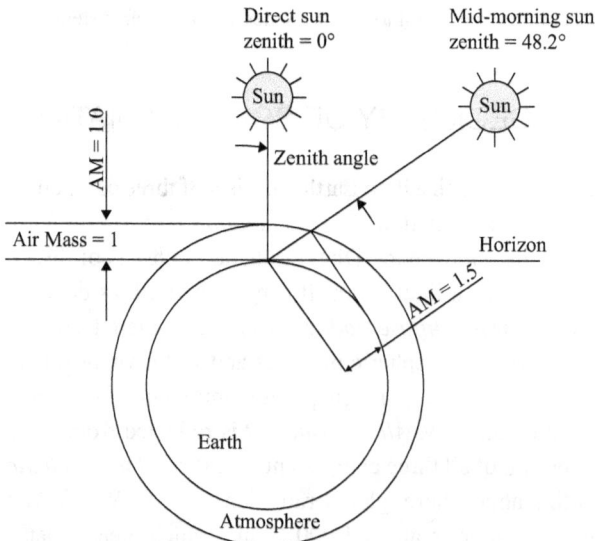

Figure 3.1. Zenith angle and AMI.

where P_a is the local atmospheric pressure and P_o is atmospheric pressure at sea level.

Global radiation (I) in W/m² is related to the AMI by the following equation:

$$I = 1367(0.7)^{(AM)} \tag{3.2}$$

Note that global radiation for AM = 1 is 959 W/m², $I = 1,370$ W/m² for AM = 0, and $I = 800$ watts/m² for AM = 1.5. We must keep in mind that this calculation is valid at the top of the atmosphere, and varies for elevation changes and atmospheric conditions. Radiation values used for testing PV modules are not exactly the same as Equation 3.2, as we will see in the following section.

3.2   SOLAR IRRADIANCE AND IRRADIATION

Another term commonly used instead of "global radiation" is *irradiance*, which is the *power* density of sunlight. Irradiance is an instantaneous quantity expressed in watts per square meter (W/m²). Irradiance for AM 1.5 is generally accepted as the standard calibration for PV cells and equal to 1,000 W/m². Figure 3.2 shows irradiance for the 24-h period of June 16, 2010, at 39° 41′N latitude, in Colorado, at an elevation of about 8,000 ft above sea level. Irradiation was measured with a horizontally mounted pyranometer, and data samples were taken every minute.

Irradiation is the *energy* density of sunlight and therefore is measured in kWh/m². Irradiation is the integral of irradiance—usually over a day.

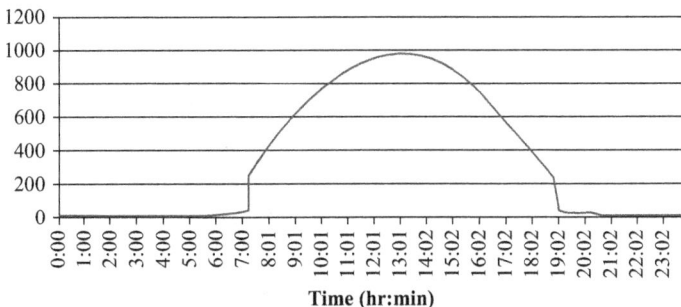

Figure 3.2. Solar irradiance. (*Source*: Courtesy of Sunnyside Solar and Ecofutures Inc., Energy Monitoring and Verification Systems.)

Figure 3.3 is the integration of irradiance in Figure 3.2 over a 24-h period. Eight thousand three hundred and forty-eight watt hour per square meter of energy was recorded at this particular location and day.

Peak sun hours (PSH) is a useful reference that represents the equivalent time (in hours) at an irradiance level of 1,000 W/m² that would equal the total irradiation for an entire day. PSH makes the integration process transparent and simplifies comparison of solar energy days at different locations. The equivalent PSH from Figure 3.2 is 8.348 h. In other words, the equivalent of the irradiance curve of Figure 3.2 would be a flat line at 1,000 W/m² for 8.348 h. Both representations equal 8,348 kWh of solar energy.

Solar radiation at a particular location and time can be predicted using formulas that represent the location of the sun. The results are approximations due to variations in the rotation of the Earth with our time standard, and the elliptical orbit of the Earth around the sun, which is not exactly 365 days. The calculations do not account for cloud cover, and ignore atmospheric variations, such as pollution and volcanic eruptions. However, the equations provide sufficient data to derive solar charts that may be used to predict solar irradiation for PV arrays at specific locations on the Earth. The equations are also useful to drive solar instrumentation designed to estimate solar radiation for a particular site. Historical data are used to estimate cloud cover and other atmospheric conditions. A combination of historical information and solar irradiance calculations offers the best prediction of solar irradiation.

The angle of the polar axis of the Earth and the plane of the Earth's orbit is called the *solar declination*. This angle is 23.45° and causes long summer sunlight hours and short winter days in the northern hemisphere. The angle of declination (δ) can be calculated using Equation 3.3.

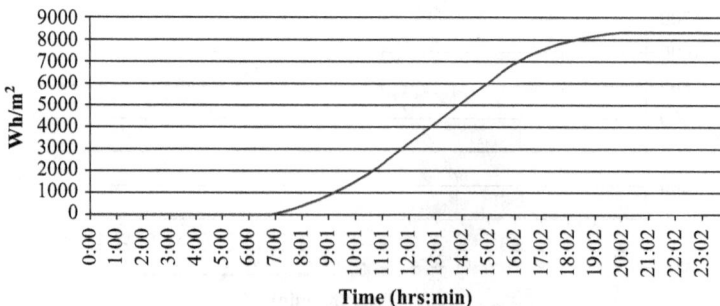

Figure 3.3. Irradiation. (*Source*: Courtesy of Sunnyside Solar and Ecofutures Inc., Energy Monitoring and Verification Systems.)

$$\delta = 23.45 \sin[360(n - 80)/365] \qquad (3.3)$$

where N is the nth day of the year, which can be determined from a Julian calendar. Figure 3.4 shows declination and how it varies throughout the year.

Since the number of days in a year is not exactly 365, and the first day of spring is not always the 80th day of the year, δ is an approximation of declination [13].

Two other coordinates are required to track the position of the sun: solar altitude and azimuth angle. *Solar altitude* (α) is the angle between the horizon and an incident solar beam on a plane determined by the zenith and the sun. As described earlier, zenith is the angle between the sun position and a vertical perpendicular to the tangent position on Earth's surface. Solar altitude is also the complement of the ζ (i.e., $\alpha + \zeta = 90°$). *Azimuth* (Ψ) is the angle between the sun's position east or west of south. $\Psi = 0°$ occurs at solar noon and Ψ is positive when the sun moves toward the east. This convention differs from a compass reading ($0° =$ north, $180° =$ south). Solar charts may also reference south as $0°$ (Figure 3.5).

The angle that describes the position of the sun from solar noon in the plane of apparent travel of the sun is the *hour angle* (ω). The hour angle is the angular displacement between noon and the desired time of the day in terms of $360°$ rotation in 24 h.

$$\omega = (12 - T) \times 360°/24 = 15 \, (12 - T)° \qquad (3.4)$$

where T is the time of day with respect to solar midnight, on a 24-h clock. For example, for $T = 0$ or 24 (midnight), $\omega = \pm180°$. For $T = 9$ a.m., $\omega = 45°$.

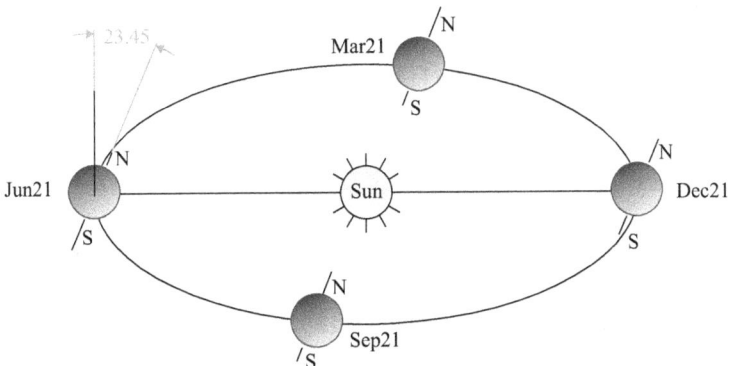

Figure 3.4. Solar declination.

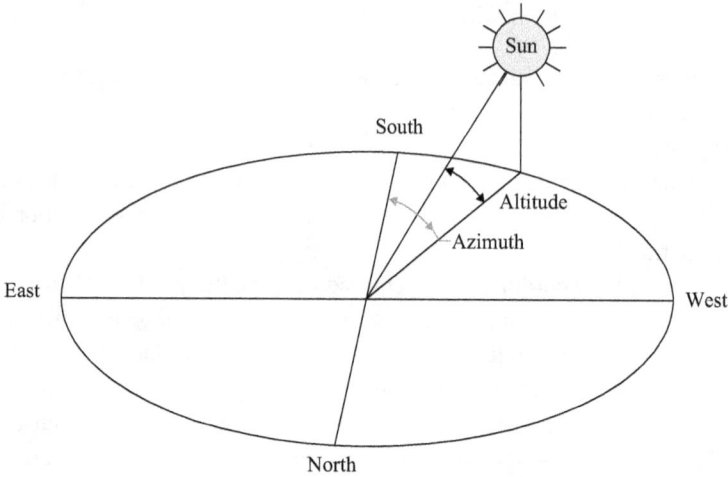

Figure 3.5. Altitude and Azimuth.

If the latitude (φ) is known, the altitude (α) and azimuth (Ψ) can be determined using the following trigonometric relationships:

$$\sin\alpha = \sin\delta\ \sin\varphi + \cos\delta\ \cos\varphi\ \cos\omega \tag{3.5}$$

$$\cos\Psi = (\sin\alpha\ \sin\varphi - \sin\delta)/\cos\alpha\ \cos\varphi \tag{3.6}$$

Solar radiation data for various locations throughout the world can be found on the National Renewable Energy Laboratory (NREL) website (http://www.nrel.gov/rredc). The data were compiled for the National Solar Radiation database from the National Weather Service from 1961 to 1990. The database contains solar data for 239 U.S. sites. A sample for one location is shown in Table 3.1. Designers of solar systems use the data to estimate the expected performance based on this historical information.

The solar radiation tables [14] show solar radiation averages for each month and a year, expressed as kWh/m²/day. The daily insolation tables also provide maximum and minimum values for each month and year, but we will use only average data. Solar radiation data are collected on fixed south-facing, flat-plate collectors, at tilts equal to latitude, latitude −15°, and latitude +15°. Single-axis and dual-axis tracker data are also provided. Single-axis trackers rotate from east to west, tracking the sun. Dual-axis trackers maintain a normal orientation to the sun throughout the day, tracking elevation and azimuth.

A limitation to the solar radiation tables is that they provide data for south-facing fixed surfaces. Other tools, such as PV Watts™, can be used

Table 3.1. Solar radiation for Boulder, CO, latitude, 40.02°

Surface orientation	Jan.	Feb.	Mar.	Apr.	May	Jun.	Jul.	Aug.	Sep.	Oct.	Nov.	Dec.	Average
Fixed array													
Lat. −15	3.8	4.6	6.1	6.2	6.6	6.6	6.6	6.3	5.9	5.1	4.0	3.5	5.4
Latitude	4.4	5.1	5.6	6.0	5.9	6.1	6.1	6.1	6.0	5.6	4.6	4.2	5.5
Lat. +15	4.8	5.3	5.6	5.6	5.2	5.2	5.3	5.5	5.8	5.7	4.8	4.5	5.3
Single-axis tracker													
Lat. −15	4.8	5.9	7.0	8.1	8.4	9.1	9.1	8.6	7.9	6.7	5.0	4.4	7.1
Latitude	5.2	6.2	7.2	8.0	8.1	8.8	8.7	8.4	7.9	7.1	5.5	4.9	7.2
Lat. +15	5.5	6.4	7.1	7.7	7.7	8.2	8.2	8.0	7.8	7.1	5.7	5.2	7.1
Dual-axis tracker													
	5.6	6.4	7.2	8.1	8.5	9.4	9.2	8.6	8.0	7.1	5.7	5.3	7.4

to predict insolation for fixed-tilt surfaces facing directions other than due south. The advantage of using solar radiation tables is that they provide a quick reference when calculating the energy a PV array will produce on an average residential rooftop.

Since power output of PV modules is rated at 1 kW/m² solar irradiance, energy output of an array of modules can be estimated simply by multiplying the table average irradiance times the size of the array in kW. Data in Table 3.1 are also the average PSH per day. For example, the 4.4 kWh/m²/day value in Table 3.1 for January at latitude and a fixed surface is also 4.4 h/day at 1 kWh/m². Therefore, a 4 kW, fixed-tilt array installed at latitude (40°) will produce 4 kW × 4.4 h/day = 17.6 kWh/day. If we apply the yearly average of 5.5 PSH for 365 days, we obtain the total estimated energy for the 4 kW array per year:

$$\text{Annual energy} = 4 \text{ kW} \times 5.5 \text{ h} \times 365 = 8{,}030 \text{ kWh/y} \qquad (3.7)$$

However, we must include derating factors, such as shading, inverter efficiency, wiring losses, and soiling. The PVWatts program by NREL provides a more practical estimate. The preceding calculations can be justified by the definition of PSH:

$$\text{PSH (h/day)} = \text{daily irradiation (kWh/m}^2\text{/day)/} \\ \text{peak sun (1 kW/m}^2) \qquad (3.8)$$

[Or PSH = watt-hours per square meter per day divided by 1,000 watt-hours per square meter]

Solar charts provide a quick method to determine azimuth and altitude versus the time of day at different latitudes. These charts are useful to evaluate a potential solar site with obstructions or shading concerns. Areas of shading caused by trees or obstructions can be identified if the height and distance from the site are known or measured. The number of hours and days of total or partial shading can be predicted using a solar chart. Figure 3.6 shows a solar chart for a location at 40° north latitude.

3.3   SHADING ANALYSIS

Even a small amount of shading will reduce the output of a PV module significantly. The electrical characteristics of shading on an array will be discussed later with PV array design. The number of hours and days of shading can be assessed by measuring the azimuth and altitude angles of the obstructions and plotting them on the solar chart. Winter months bring

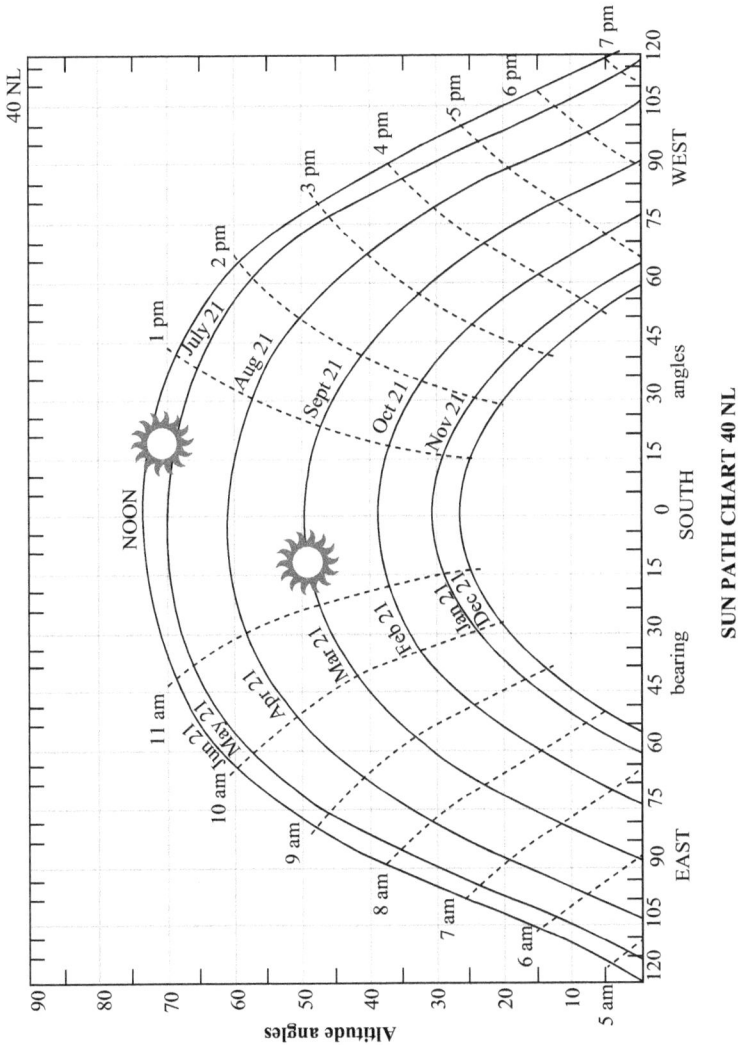

Figure 3.6. Sun Path Chart for 40° North Latitude. (*Source:* Courtesy of Solar Energy International (SEI) [15].)

low sun altitudes, and distant obstructions can significantly reduce winter performance. Power lines and overhead communication wiring will also cause significant reduction in PV output.

Several tools are available to estimate shading. Solar Pathfinder™ is a mechanical device that projects a plot of the surroundings from potential PV site on a screen. It compares the plot with solar chart data and displays the areas and times of potential shading. The pathfinder also calculates the percentage of solar energy that reaches the site at different times of the day and months of the year. Online technologies such as Google Earth and SketchUp provide shading estimates using satellite imaging and historic solar radiation data. New smart phone applications are also available that are capable of measuring shading, calculating solar irradiation, and integrating historical data to produce estimates of solar production. The phone applications use compass and mapping functions with the camera to develop shading charts and calculations of annual solar irradiation. A sample shading report is available in Appendix A.

Solar Pathfinder is the classic mechanical tool that projects trees, buildings, and other objects that cast shadows on the planned PV array. It uses a plastic dome that reflects a shading pattern on a solar chart, which is specific to latitude. The chart is marked with percentage shading factors for each half-hour during the solar day. These factors are added for each month to give a total shading percentage. Annual shading percentage is the sum of the monthly percentages (Figure 3.7).

The Pathfinder includes a built-in compass and bubble level, and can be set up at the planned PV location with extendable tripod legs and an adjustable base. Measurements should be taken at several locations where PV modules may be subject to potential shading. Two types of charts are supplied with the Pathfinder: the sun-path chart and an angle estimator that gives azimuth and elevation of any object reflected in the dome. The sun-path chart example in Figure 3.8 provides the information necessary to estimate the percentage of energy lost due to shading.

The Pathfinder must be adjusted for proper magnetic declination. The center chart tab can be rotated to align the chart for proper magnetic declination, and the resulting chart is aligned for true north orientation. The sun-path chart in Figure 3.8 is designed for 40° north latitude, and is adjusted for magnetic declination of 10° east of true north (see the *Magnetic Declination* section later).

The sun-path diagram of Figure 3.8 shows 12 horizontal arcs—one for each month, as well as vertical lines or rays for solar time. Each arc shows the mean sun's path across the sky for that month. The sun-path arc for June is closest to the center, and the sun-path arc for December is farthest from

Figure 3.7. Solar pathfinder.

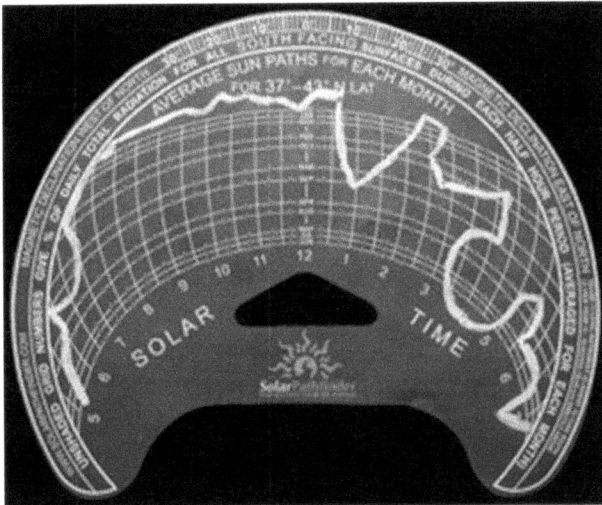

Figure 3.8. Pathfinder chart for 40° north latitude solar site.

the center. Solid radial lines represent each half-hour. Solar time is approximately standard time at the center of each time zone.

A disadvantage of the Pathfinder is that it is difficult to use under direct sunlight. An overcast day provides the best shading projections on the chart. It is also somewhat difficult to trace the shadow outline while inserting the wax pen through the narrow side opening. It is possible to take a digital picture of the chart instead of tracing the shadow outlines with a wax pen.

Newer electronic Pathfinder tools record shading and perform all calculations without the mechanical chart and pen method. However, it is helpful to follow the mechanical process to obtain a good understanding of the method the new equipment makes transparent.

Figure 3.8 is a Pathfinder chart that reflects significant shading objects in the afternoon hours, particularly around 1 p.m. By adding the shading factors along the January curve, we get a potential available radiation percentage of 64% (2 + 3 + 4 + 5 + 6 + 7 + 7 + 8 + 8 + 8 + 6). In other words, 36% of potential radiation is blocked out during the winter month of January.

3.4   MAGNETIC DECLINATION

True north and magnetic north are not the same except at certain locations in the world. Figure 3.9 shows the difference between true north and magnetic north in the United States, which is called the *magnetic declination*. Declination is zero approximately along the 90° west longitude, which is called the *agonic line*. Declination is negative east of this line and positive west of this line. Boulder, Colorado, is approximately 10° east of north in declination. That means a compass will point 10° east of true north. Therefore, we must subtract 10° from our compass reading to obtain true north.

Magnetic north is created by the molten metallic outer core of the Earth, which is about 3,000 km below the Earth's surface. The poles of this field do not align with the axis of rotation of the Earth (true north). Magnetic

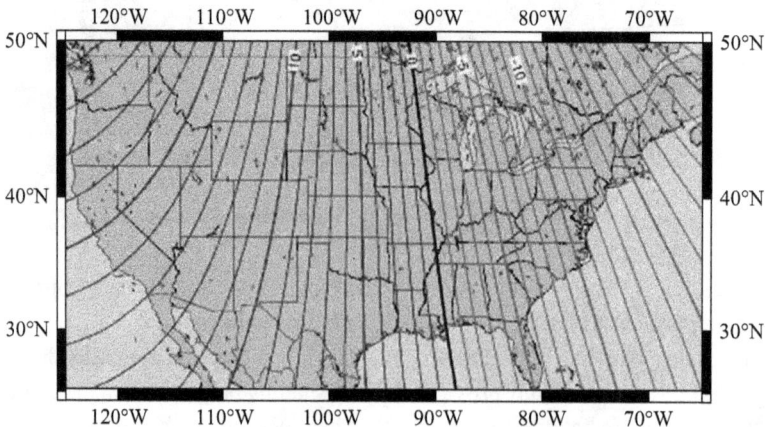

Figure 3.9. Magnetic declination.

declination does not remain constant with time, because of the fluid characteristic of the molten core. Maps showing declination may be out of date, and current declination values can be obtained for a particular latitude and longitude from a government website (e.g., http://www.ngdc.noaa.gov).

3.5 ARRAY ORIENTATION

The ideal orientation for a PV array is perpendicular to the sun from sun up to sun down. Two-axis trackers can be used to maximize the harvest from the sun. Single-axis trackers also improved energy harvest as compared to fixed arrays. However, it is more economical and space-advantageous to install modules with a fixed orientation. Trackers have high installation costs due to motors, gearing, and controllers. If several trackers and towers are installed, more land space is required to prevent shading. They also require maintenance. Fixed arrays have no moving parts, are easier to protect from high winds, are aesthetically more appealing, and still capture a significant portion of total radiation.

The best year-round energy harvest orientation for a fixed PV array is azimuth of 0° (south) and a tilt angle equal to the latitude. Data collected by NREL (http://www.nrel.gov/rredc) and Sandia National Laboratory confirm this general rule. However, it may not be practical to mount the array facing directly south or at a tilt angle equal to the latitude. Tabulated data from these solar reports provide radiation values for deviations from optimal azimuth and tilt angles. This information may also be obtained using the PVWatts calculator provided by NREL on its website (http://www.nrel.gov/rredc/pvwatts).

The PVWatts program may be used to determine annual energy production of a PV array at various orientations. Table 3.2 was derived from PVWatts data by entering various tilt and azimuth orientations for a 1 kW array located at latitude 39.6 north and longitude 104.9 west. Default conditions are 180° (south) orientation and tilt of 40°. Default input results in an annual energy output of 1,453 kWh with a 1 kW array, which is selected as 100% of optimum. Table 3.2 offers a quick reference when evaluating the effect of various roof orientations. Renewable Energy Credits are usually based on predicted annual energy output, and reductions of the rebate are applied when less than 90% of optimum annual output is caused by the orientation. For example, an east orientation (90°) at a 20° tilt will result in 82.9% (1,205 kWh) of optimum output.

Further analysis using PVWatts shows that actual maximum energy output occurs at a 38° to 39° tilt and 165° to 170° azimuth, which results

Table 3.2. Energy and orientation

Azimuth	Tilt (degrees)					
	10 (%)	20 (%)	30 (%)	40 (%)	50 (%)	60 (%)
90	83.2	82.9	81.3	78.6	74.9	70.3
120	87.4	90.8	92.2	91.6	89.1	84.9
150	90.0	95.6	98.6	99.2	97.5	93.5
180	90.5	96.4	99.4	100.0	98.2	94.2
210	88.6	92.8	94.8	94.3	91.9	87.5
240	84.8	86.0	85.5	83.5	80.0	75.1
270	80.1	77.3	73.6	68.8	64.6	59.7

Source: Courtesy of Michael Coe.

in 1,465 kWh of energy per year. For default inputs other than tilt and azimuth, see Appendix A—Derating Factors.

3.6 SITE ANALYSIS

The first step to a good solar installation is the site analysis. What is the available solar irradiation at the site? Is a good southern exposure available without significant shading? Are the roof angle, surface area, and orientation adequate to capture a significant amount of the radiation? What are the economics and incentives available for the customer? These are a few of the questions that must be addressed before moving forward on a solar installation.

These questions are answered with a site visit and evaluation. Orientation and shading issues can be evaluated manually with geometric and trigonometric tools provided at the beginning of this chapter. A tape measure, compass, and inclinometer are the tools necessary to plot the PV array orientation and shading obstructions. Measure the distance, height, and outline of trees, buildings, and all possible shading objects from the proposed location of the PV array. For a large array, these measurements should be made from several locations where the lowest modules may be mounted. The solar chart for the appropriate latitude will provide a good approximation of solar angles and hours of solar radiation. By plotting shading objects on the solar chart, estimates of shading time and array coverage are calculated. For each month of the year, determine total available irradiation and estimate the percentage of shading times. The result is summed into an annual prediction of energy output for the array at optimal tilt and orientation. The next step is to determine the actual tilt and orientation of the array and reduce the predicted output by an appropriate derating factor. The final calculation of predicted output incorporates the efficiency of the modules, the inverter efficiency, and a derating factor for all wiring.

A universal tool most solar designers and installers use is the PVWatts program available online from NREL (see http://www.nrel.gov/rredc/pvwatts). The PVWatts program provides available solar radiation data based on historical data. Program inputs are latitude and longitude, tilt angle, orientation, and derating factors for equipment and wiring. Additionally, the designer must lower the prediction if there are shading issues. The program output is conservative, and may be increased by changing the default derating factors.

Many other tools beyond those mentioned earlier (Solar Pathfinder, Google Earth, and SketchUp) are available to assist with site evaluation and integration of shading information. The SunEye™ by Solmetric will electronically evaluate a site and generate an accurate energy evaluation of the potential site. Solmetric also developed an application (app) for the iPhone® that is much less expensive, but does not have the accuracy of the SunEye or other similar instruments. The iPhone app (Solmetric iPV) relies on the camera function, compass (3GS), and inclinometer built into the iPhone. The user manually scans the horizon and obstructions, using the camera function as a siting tool and the inclinometer to measure the object's angle above the horizon (object-altitude angle). A digital recording is made of the object-altitude angles versus the azimuth angle. This recording is the shade outline, which is then plotted on the solar chart for the site location. Solar chart data are obtained automatically by the program by using a GPS and looking up solar radiation information from the nearest weather station. The program analyzes these data and creates a monthly and annual report of predicted energy output for the array. The array may be fixed, one-axis, or dual-axis. The program also incorporates efficiency and capacity information for many types of inverters and PV panels. A sample report from the iPhone application is provided in Appendix A.

PROBLEMS

Problem 3.1

Calculate the number of hours the sun was above the horizon today at your location.

Problem 3.2

Using data from a pyranometer, plot the irradiance curve for yesterday at the pyranometer location. Calculate irradiation using digital integration.

Problem 3.3

What are the equivalent PSH for the irradiation calculation in Problem 3.2?

Problem 3.4

Use the solar chart for 40° north latitude, and estimate the solar altitude and azimuth on February 21 at 10 a.m. What are the altitude and azimuth angles for September 15 at 10 am?

Problem 3.5

Calculate solar altitude and azimuth for 40° north latitude on February 21 at 10 a.m. What is the percentage deviation from the calculated values and the solar chart values determined in Problem 3.4?

Problem 3.6

Design a PV system for a commercial building with a flat roof shown in Figure 3.10. The roof has a parapet around it that is 3 ft above the roof deck. An air conditioner is located, as shown, 4 ft above the roof and 5 ft square.

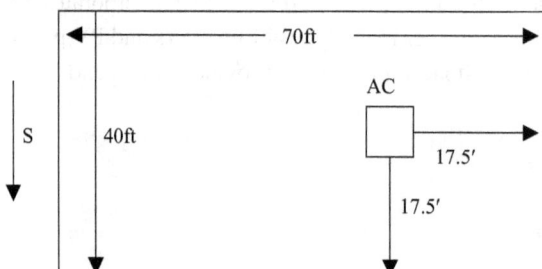

Figure 3.10 Roof layout.

The total roof area is 70 ft by 40 ft and located at 40° north latitude. PV modules are to be installed at a tilt angle of 25°, in portrait mode, facing due south. The bottom of the modules is 6 in above the roof deck. Determine the maximum number of modules that can be installed on the roof so that no module will be shaded for 6 h every day of the year. Pay attention to shading from the AC unit, the parapet, and other modules.

CHAPTER 4

INTERCONNECTED PHOTOVOLTAIC SYSTEMS

Interconnected photovoltaic (PV) systems have become the most economical application of renewable energy power systems. "Grid-connected" (GC) and "utility-interactive" are other terms used to describe renewable energy systems interconnected to the utility power grid. Before incentives and mass production, PV applications were limited to remote installations where utility power was not available and the load was small. PV systems were also used for satellite and space applications where cost was not a deciding factor. Subsidies in the form of utility rebates and investment tax credits have driven costs down and provided investment incentive for residential and small commercial applications. Technical problems with many distributed generation systems connected to the utility grid have been solved or mitigated. Increased demand has lowered the cost of PV modules, and balance-of-system equipment costs have also come down due to scaled-up manufacturing. Net metering and public utility regulatory laws have encouraged small generation systems. Since many small PV systems are now being installed in the United States, this chapter will explore system design and operation of PV as an interconnected system.

System reliability has risen significantly in the last 10 y. In the 1970s and 1980s, solar and wind systems had the rap of unreliability. Utility-interconnected inverters failed frequently, caused harmonic distortion, and created a poor power factor. It was true that components of early inverters were not as reliable as today's semiconductor devices. Early designs also did not filter distortion adequately. Today, inverter manufacturers offer 10 y warranties, 97% efficiencies, power factors of 99%, and less than 4% total harmonic distortion (THD). PV efficiency has increased to around 14% for polycrystalline modules and 20% for single-crystalline, double-sided high-performance modules. PV module performance warranties are now provided for 20 to 25 y.

GC utility-interactive systems are designed to deliver maximum PV power directly into the utility power grid. Some systems incorporate battery backup in addition to the utility interconnection, but the hybrid systems are not as cost effective and not discussed in this chapter. Storage systems are still an expensive component and require regular maintenance. The beauty of a simple GC system is the maintenance-free aspect—no moving parts or electrochemical concerns. The utility grid serves as the storage medium.

4.1   INVERTERS

The heart of any direct current renewable energy system is the inverter. *Stand-alone inverters* do not connect with the utility grid but produce alternating current (AC) power at a set frequency and voltage. Stand-alone inverters are designed to provide AC power when utility service is not available or economical. The *GC inverter* is designed to convert direct current (DC) voltage and current to AC voltage and current that is in synchronism with the utility frequency, voltage, and current. A GC inverter is also called a "line-commutated inverter" because it is designed to synchronize with the grid. There are many types and designs of inverters used for renewable energy systems. This chapter will discuss the general construction and operation of the GC inverter used for PV systems. Basic designs of any inverter are the square wave, sine wave, modified sine wave, pulse-width-modulated (PWM), high-frequency, and steering bridge. The type of design determines the quality of AC output as described later in this chapter. These designs are incorporated in the GC inverter and stand-alone inverter. GC inverters are also classified by size and application.

The *string inverter* is used in residential and small commercial applications and sized for PV arrays in the 1 to 12 kW range. It is called a string inverter because the input source consists of one or more "strings" of series-connected modules. The inverter input voltage range determines the number of modules in each string. String voltage is also limited by the National Electrical Code (NEC) requirements for residential and commercial construction—600 or 1,000 V maximum. Strings are connected in parallel to increase the array size and source power to the inverter. All modules in the array must have the same orientation and be without shading issues for optimal performance. String inverters are often installed in parallel to increase power beyond the 12 kW range.

Microinverters, also called "module-level inverters," are sized to individual PV modules (200–300 W). Microinverters may be integrated into the PV module frame or installed separately under the modules on the racking equipment. Since each microinverter produces AC current from the individual module, they are connected in parallel to a dedicated AC branch circuit breaker in the electrical distribution panel. Advantages of this type of inverter are that the PV array modules are not restricted to a single orientation, the array is less susceptible to shading problems, and microinverters also claim increased reliability and performance through redundancy. If one PV module or inverter fails, the remaining modules continue to operate. A *central inverter* is used in commercial applications where the array size is from 30 to 500 kW. The PV array is an assembly of many strings of modules, which are connected in parallel in a combiner enclosure. The combiner also contains fuse protection for each string. Several combiner circuits (source circuits) may also be connected in parallel to match the total array source power to the rated inverter input power. Array modules for a central inverter must be homogeneous—have equivalent ratings and the same tilt and orientation.

Utility scale inverters are used in PV power production systems rated 500 kW-1 MW. This size of inverter is very similar in general design and operation as central and string inverters, but must include components rated at much higher voltages and currents. They also incorporate algorithms and control functions that ensure that the output does not cause problems on the utility grid. For example, stability requirements may necessitate a fault ride-through function and voltage adjustments.

For safety reasons, the GC inverter must disconnect from the utility whenever utility voltage is not available or is not within a nominal range. Underwriter Laboratories (UL) 1471 is the standard developed by Underwriters Laboratories that inverters must meet to be acceptable by the NEC. For many years, utilities required that a manual safety disconnect be installed between the grid and inverter that would ensure that the inverter would not continue to generate voltage on the grid when a utility power failure occurred. Electricians and linemen insisted on the visible lockable switch for their safety. In the last few years, this requirement has been dropped, because the GC inverter has proven to meet UL, NEC, and Institute of Electrical and Electronics Engineers (IEEE) standards requiring the inverter to shut down within seconds of loss of utility voltage or unstable frequency conditions.

IEEE Standard 929-2000 describes the recommended practices of utility interface of residential and intermediate PV systems. The standard was revised in 2003 by the IEEE Standards Coordinating Committee 21. IEEE 1547 is the functional standard that defines how an inverter must disconnect from the grid when the grid de-energizes or is in abnormal condition. The standard also defines voltage regulation, grounding, synchronization, and disconnect means for the GC inverter. Nominal AC voltages for residential and commercial applications may be 120, 208, 240, 277, or 480 V. Clearing times (i.e., time to disconnect) are listed in standards by percentage of the "nominal voltage" rating of the inverter (see Table 4.1).

The GC inverter must also disconnect when frequency varies from 60 hertz (Hz). Clearing time is 0.16 s when frequency drifts above 60.5 Hz or below 59.3 Hz. For inverters above 30 kW, the frequency range must be adjustable to coordinate with utility protective equipment (i.e., relays). If the inverter disconnects for a voltage or frequency disturbance, it must monitor voltage and frequency for 5 min after restoration and before reconnecting to the grid.

PV modules are designed to operate at specified voltages and currents. Each silicon cell generates approximately 0.6 V DC. When PV modules were used mostly for charging 12 V batteries, 36-cell modules were the norm, producing 14–20 V DC. As the demand grew for utility-interconnected systems, 54- to 72-cell modules became more common. PV panel voltages range from 30 to 45 V and are wired in series strings to generate DC voltages of 300 to 600. This high voltage is more efficient in PV array design because it reduces the current and allows smaller wiring from the modules to the inverter. High voltage is also necessary with GC inverters that connect to 240 V or higher utility voltages. DC voltage must be higher than AC voltages to cause power to flow from the DC system to the AC system. This will become apparent as we discuss AC to DC rectifiers and inverter electronic operation later in this section.

The *square wave inverter* is the least costly and has the best surge capacity. However, the high distortion in output waveform makes it unacceptable for most applications. IEEE Standard 1547–2003 requires that any source connected to a utility line must have less than 5% THD. *THD*

Table 4.1. Inverter clearing times

Voltage range (%)	Clearing time (s)
<50	0.16
50–88	2.00
110–120	1.00
>120	0.16

is the ratio (as a percentage) of the sum of root-mean-square (rms) voltage (V) values of all harmonics above the fundamental frequency voltage (V_i) and the rms value of the fundamental frequency (i.e., 60 Hz in the United States) voltage (V_0).

$$\text{THD (\%)} = 100 \times \Sigma\, V_i/V_0 \qquad (4.1)$$

The basic construction of a square wave inverter is shown in Figure 4.1. The DC source shown may be a PV array, DC generator, or stand-alone battery. Solid-state switches Q1 through Q4 may be silicon-controlled rectifiers (SCRs), insulated gate bipolar transistors (IGBTs), or power metal-oxide field-effect transistors (MOSFETs). IGBTs are used for most high-power inverters, but SCRs and MOSFETs are still common in low-power and low-voltage systems.

To generate a square wave from the DC input, Q1 and Q3 are turned on at the same time by a gate control circuit (not shown) for the first half of the cycle. Q1 and Q3 are turned off at the beginning of the second half cycle, and Q2 and Q4 are turned on. The output (V_o) may be a discrete load or utility-grid connection. This simple control system creates a square wave, as shown in Figure 4.2. A sine wave is plotted with the square wave to show what the ideal output of an inverter should replicate. *Sine wave*

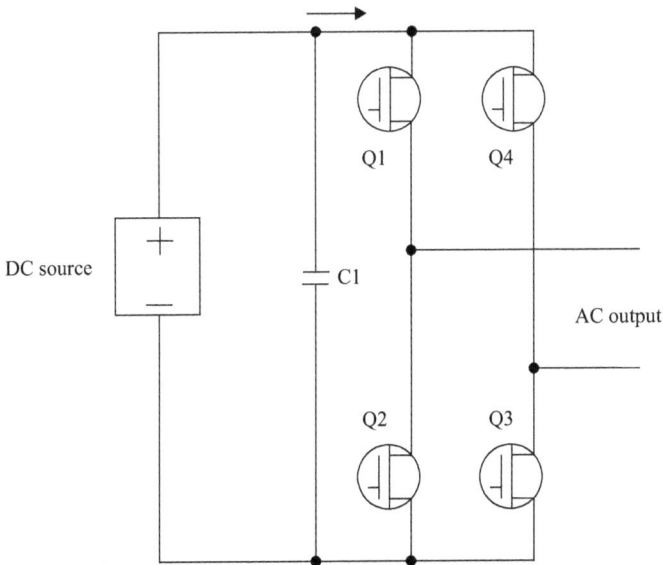

Figure 4.1. Simple square wave inverter.

Square wave inverter output

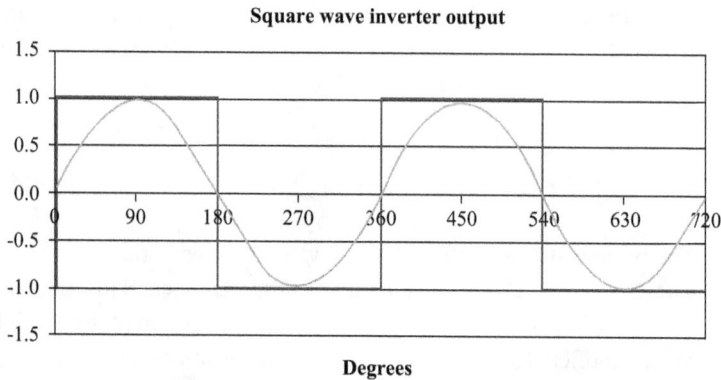

Degrees

Figure 4.2. Inverter square wave versus the sinusoidal waveform.

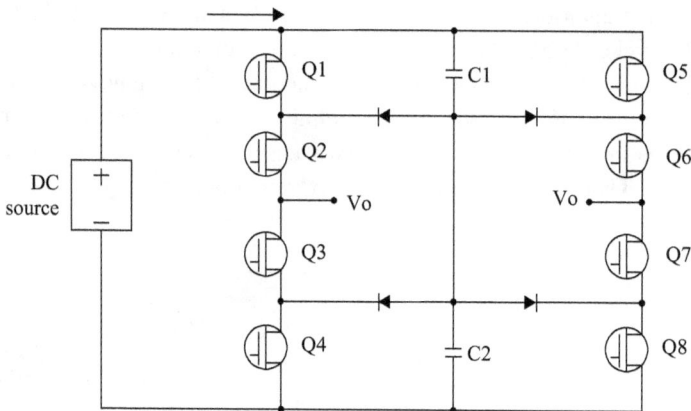

Figure 4.3. H-bridge inverter circuit.

inverters produce output with less DC distortion and harmonic content than the simple square wave inverter. However, the electronics are more complicated and include capacitive and inductive filtering.

A *modified sine wave inverter* is similar in construction and operation to the square wave inverter. Four solid-state switches and a capacitive voltage divider are added to the square wave design, as shown in Figure 4.3. Gate controls turn on the appropriate solid-state switches to create a stepped voltage waveform, as shown in Figure 4.4. This particular type of modified sine wave inverter is a five-level H-bridge. H-bridges can be designed with $4n + 1$ levels, increasing the approximation to a sine wave and decreasing distortion at higher levels. A five-level H-bridge as shown may be used to

Modified sine wave inverter output

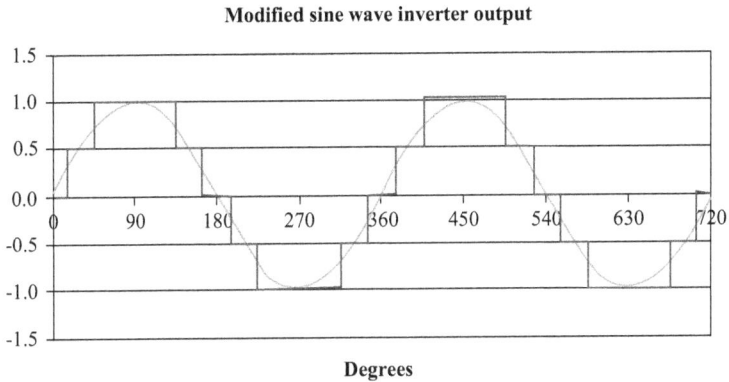

Degrees

Figure 4.4. Modified sine wave and pure sinusoidal wave.

power a computer, but it does not meet the 5% distortion standard required for interconnections.

The *PWM inverter* can produce sinusoidal waveforms of fixed amplitude and frequency. With the addition of line filtering, the output waveform can closely match the utility frequency, voltage, and sinusoidal wave shape. PWM inverters do this by controlling the on/off times of the solid-state switches with synchronized gate controllers. The duty cycle (on/off ratio) of the pulses determines waveform shape, frequency, and amplitude. It is capable of producing waveforms with an average value of zero (i.e., no DC component). This characteristic is very important to motors on the utility system. Zero average value means no DC current losses in the motor windings and high efficiency. Output distortion for the PWM inverter is less than the 5% interconnection rule, and most PWM inverters specify 4% or less THD.

Figure 4.5 represents a PWM waveform created from a triangular carrier waveform and a reference sinusoidal waveform, as shown in Figure 4.6. Switching devices, as shown in Figure 4.1, are controlled by the PWM signal to generate a sinusoidal output that can be synchronized with the utility grid. The reference signal can be phase-shifted with respect to the grid voltage, and the power factor can be controlled by shifting the output current phase relationship. For this reason and low-harmonic generation, most inverters are designed with PWM.

GC inverters may incorporate a high-frequency transformer to provide isolation between the DC PV input and the utility grid. The transformer also steps up the voltage and reduces harmonics to improve overall efficiency. The particular inverter shown in Figure 4.7 also incorporates a "steering bridge." The input inverter switches (Q1 through Q4) create a square wave

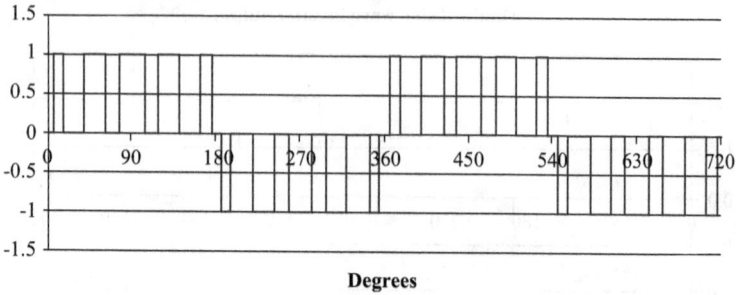

Figure 4.5. Pulse width modulated waveform.

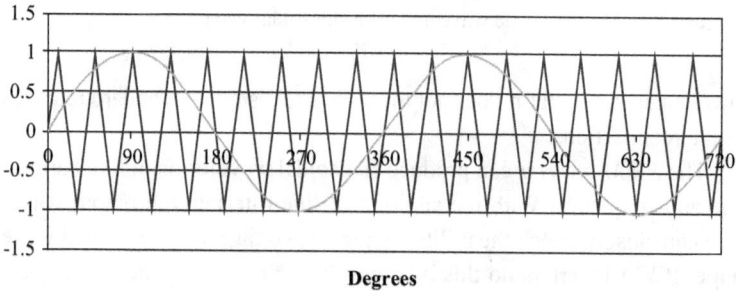

Figure 4.6. Triangular carrier waveform and sinusoidal reference.

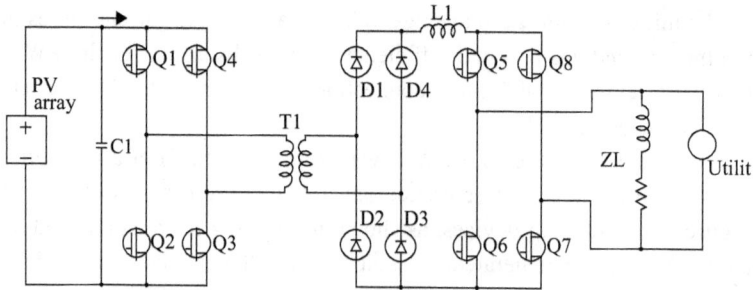

Figure 4.7. Typical grid-connected inverter circuit.

or modified square wave, which is stepped up by transformer T1 and again rectified by diodes D1–D4. The output bridge (Q5–Q8) "steers" the recti-fied current into a full sine wave that flows into the utility grid. The output bridge switches at 100–120 Hz and at zero crossing to make the system more efficient than higher speed switching circuits.

4.2 PHOTOVOLTAIC CELLS

Extracting electrical power from sunlight, like so many other sources of energy, is an inefficient process. Traditional PV technology uses layers of semiconductor material. Under normal atmospheric conditions, only about 18 to 20% of available energy is converted to electrical energy. This means that large areas or arrays of panels are necessary to extract the amount of power needed to accommodate modern electrical demands.

One of the main factors limiting efficiency of PV technology is the nature of sunlight and the ability of PV cell material to absorb all available energy. Sunlight comprises a broad spectrum of electromagnetic radiation, including infrared, visible, and ultraviolet light. Each region of this spectrum has a range of wavelengths: infrared has the longest, ultraviolet the shortest, and visible light having wavelengths between the two. Along the electromagnetic spectrum, the short wavelengths have the most energy. Unfortunately, the most common PV materials, like silicon, do not absorb solar radiation at the shorter wavelengths as well as other materials.

4.2.1 SILICON CELL TECHNOLOGY

The solar cell is the core of any PV system. The PV cell is similar to the p–n junction used in diodes and many semiconductor devices. Many types of semiconductor materials are used to create a PV cell, but silicon is presently the most common one. Silicon dioxide is very abundant, making up over half of the Earth's crust. However, the pure silicon needed to create efficient cells is very energy intensive to produce. About 50 kWh/kg of energy is needed to produce silicon of 99% purity. High-efficiency cells require a purity of one part in 10^8, and over 200 kWh/kg is used to obtain this purity. Arc furnaces are required to separate silicon from oxygen, and ingots of silicon crystals are grown from the pure silicon. The ingots of single-crystal or multicrystalline silicon are then sawed into wafers to create discrete cells (Figure 4.8).

Single-crystal (monocrystalline) cells are the most efficient (18% to 20%), but require the most energy to manufacture. Polycrystalline (multicrystalline) cells require less energy to produce and are less efficient (16% to 18%). Amorphous silicon cells take the least energy to manufacture, but are even less efficient (8% to 10%). Another approach involves the deposition of thin layers of amorphous silicon cells on stainless-steel ribbon. An advantage of amorphous silicon cells is that they have a much larger absorption constant than crystalline cells, and may be used in thin-film

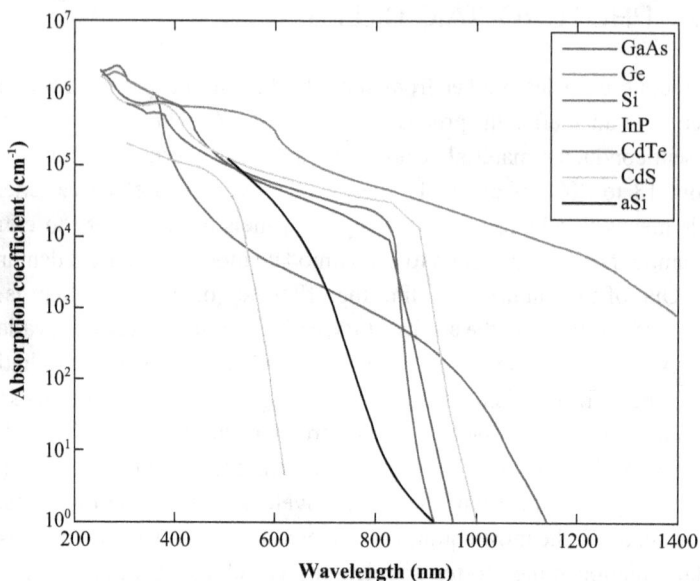

Figure 4.8. Absorption coefficients [16].

technology. Thin-film technology may also use other materials, including cadmium telluride (CdTe) and copper indium gallium diselenide (CIGS). Cells of these materials are deposited on either glass or stainless steel substrates.

In any case, the growing and sawing of ingots is a highly energy-intensive process, and carbon dioxide is a byproduct of the process. Fortunately, the greenhouse gas generated by the process is significantly less than that when fossil fuels are used to produce the same energy over the lifetime of the PV cells. Also, the CO_2 produced may be sequestered and kept out of the atmosphere using newly developed techniques of capturing the greenhouse gas. Details of solar cell fabrication and manufacturing processes are presented very thoroughly in publications referenced at the end of this chapter [17].

Recent research has established grid-tied energy payback time values for several PV module technologies (see Table 4.2). The energy required for producing PV modules with any of the technologies of Table 4.2 does not exceed 10% of the total energy generated by the system during its anticipated operational lifetime. Data assume a 30-y period of performance and 80% maximum rated power at the end of equipment lifetime. Table 7.1 assumes 1,700 kWh/m²/y solar radiation and a 75% performance ratio for the system compared to the module [18].

Table 4.2. PV production energy cost

Cell technology	Energy payback time (y)	Energy used to produce modules compared to total generated energy (%)
Single-crystal silicon	2.7	10.0
Nonribbon multicrystalline silicon	2.2	8.1
Ribbon multicrystalline silicon	1.7	6.3
Cadmium telluride	1.0	3.7

4.2.2 OTHER PHOTOVOLTAIC CELL TECHNOLOGIES

A new technology is currently being developed at Stanford University that increases PV module efficiency to 50%. Photon-enhanced thermionic emission (PETE) uses both light and heat from the module to generate electricity. Existing technology becomes less efficient as cell temperature rises, but PETE captures wasted heat with a thermal converter. The semiconductor material is coated with a thin layer of cesium. PETE works well in a concentrator array where temperatures reach 200°C. Researchers are working on economic production of PETE.

Stacking cells to increase PV module efficiency has been used by several manufacturers. Cells of different materials are stacked to form multiple junctions. Each layer absorbs energy from a different wavelength band, and the layers are connected in series with one another. Efficiencies close to 40% have been measured, but the manufacturing cost of this technology has prevented widespread use.

4.3 PHOTOVOLTAIC MODULES

The PV module in an interconnected system is designed to generate voltage, current, and power at levels that can be used with an inverter to connect to the utility grid. As stated in the previous section, array voltage of 300 to 600 V DC is a common design output to match inverter input and grid voltage output. Current is minimized to lower wire losses and reduce wire size. Power output for an array is dependent upon many factors, including the area available for modules, the load usage at the site, utility limits, service panel limits, and investment cost. This section briefly describes the construction of PV modules and how the PV modules are connected in a grid-interconnected system.

Typical current–voltage (*I–V*) characteristics of a PV module are shown in Figure 4.9 for a low-voltage module (12 V DC nominal) with 36 cells [19]. To create the *I–V* curves, electrical tests are performed on the module under standard test conditions (STC) of 25°C, air mass index of 1.5 (AMI 1.5), and different intensities of solar radiation. Load resistance is varied to obtain output power from short circuit to open circuit. Current output is directly proportional to the irradiance to the module. Voltage, however, is not changed appreciably by irradiance, and is limited by the maximum cell voltage of about 0.6 V.

The characteristic equation for an ideal PV cell is given in Equation 4.2:

$$I = I_L - I_0 \left[e^{(qV/kT)} - 1 \right] \tag{4.2}$$

where,

I_L is the component of cell current due to photon radiation

$q = 1.6 - 10^{-19}$ C

$k = 1.38 - 10^{-23}$ J/K

T is the absolute temperature (°K)

Under short-circuit conditions, voltage is zero, and Equation 4.2 yields $I = I_L$, which is the short-circuit current (I_{sc}). Short-circuit current is always specified for a module under STC, and is used to calculate maximum system current for module strings and PV arrays.

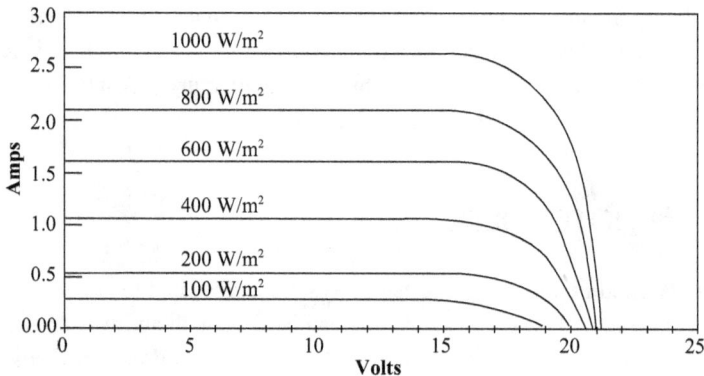

Figure 4.9. Typical *I–V* curve for a silicone PV module.

The *Kelly cosine function* is a useful tool for estimating the current output of a PV cell or module given the zenith angle—the angle between the direct sun and the normal angle (perpendicular) to the module.

$$I = I_o \cos \theta \qquad (4.3)$$

where, I_o is the maximum current output with normal light incidence (zenith angle $\theta = 0$).

The Kelly cosine function is accurate up to about 50°.

Temperature curves may not be included with module specifications, but the temperature–voltage relationship is usually specified as a coefficient. As ambient and cell temperature rises, the voltage of each cell and the module decreases, as shown in Figure 4.10. The voltage decrease explains the decrease in power performance of the PV module as temperatures rise. Conversely, voltage increases as the temperature decreases below 25°C, and the maximum design voltage is calculated for the lowest expected temperature at the site location. Generally, module efficiency drops 0.5%/°C.

Cell temperature may be estimated with a linear approximation based on ambient temperature, irradiance, and a nominal open-circuit temperature. A nominal operating cell temperature (NOCT) is determined when the cell is operated at open circuit, ambient temperature of 20°C, AMI of 1.5, 0.8 kW/m², and a wind speed of less than 1 m/s.

$$T_c = T_a + (\text{NOCT} - 20) \times G/0.8 \qquad (4.4)$$

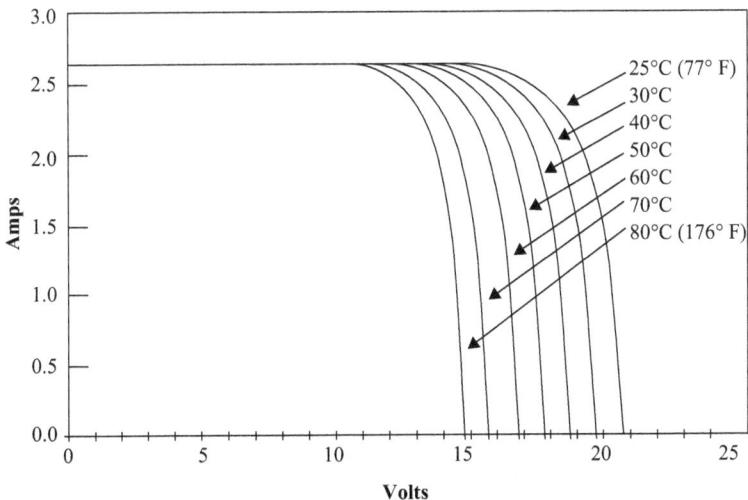

Figure 4.10. Temperature *I–V* curves for silicon PV modules.

where,

T_a is ambient temperature

T_c is cell temperature

G is the irradiance for expected ranges of operation

The open-circuit temperature coefficient for an ideal silicon cell is typically −2.3 mV/°C. However, module construction and cell type cause significant deviations from this value. A polycrystalline module specifies coefficients around 0.12 V/°C, while a high performance monocrystalline module specifies coefficients of 0.17 V/°C. Since voltage is the dominant factor in power output, system design must consider methods to reduce operating temperatures. High temperatures also decrease life expectancy of PV modules. The design must also consider high voltages at very low temperatures and safety margins for insulation.

High-performance modules use a larger number of cells to lower the voltage drop caused by high temperatures. Another design method to minimize temperature effects is to incorporate amorphous cells around single-crystal cells to increase voltage. Airflow is critical under the modules to keep the cells cool. Standoffs and roof spacing can increase airflow and help improve voltage during high temperatures.

Peak power tracking functions are incorporated into inverter designs to maximize power at all levels of irradiance and temperatures. Maximum power is generated by a PV module when the product of voltage and current is at a maximum. As shown on the $I–V$ curve for the low-voltage module, this point is near the "knee" of the curve. To the left of the knee, current is high but voltage is low, and to the right of the knee, the voltage approaches maximum open-circuit voltage but the current drops off quickly. As temperature changes, or irradiance levels change, the knee moves and the inverter must change loading to track the maximum power point (MPP). This process is called *maximum power point tracking* (MPPT).

GC inverters and stand-alone inverters use MPPT to match loads with PV array output. The stand-alone inverter MPPT is designed to generate enough power to meet the connected loads and any battery charging requirements. The PWM control circuit monitors the AC output load and varies the duty cycle of the control pulse to increase or decrease PV output accordingly. The PV array may not always be at the MPP if the connected load falls below the PV array capacity. GC inverters are designed to produce maximum power at all times because the utility interconnection appears as an infinite load to the inverter. Therefore, the MPPT control circuit monitors the output of the inverter and varies the duty cycle of the PWM pulse to maximize output at all times. The control circuit compares

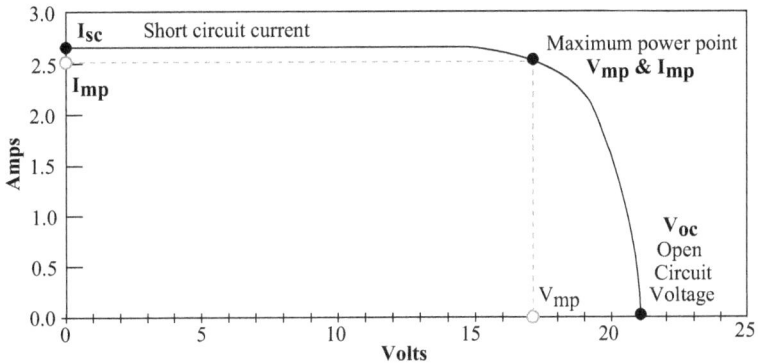

Figure 4.11. MPPT and the IV curve for a PV module.

an increase in the duty cycle of the PWM and the corresponding increase to the power output. If the differential is negative, the duty cycle is decreased. A positive differential increases the duty cycle. This process is repeated quickly and frequently to ensure that the inverter responds to light-intensity variations (e.g., moving cloud cover). MPPT also responds to PV temperature variations, which change the I–V curve characteristics (Figure 4.11).

Shading is a significant cause of power loss for a module, string, and the entire PV array. Figure 4.12 shows the extreme effects of shading on the performance of a particular low-voltage module.

With only one cell out of the 36 total module cells shaded, the module power output will decrease 75%. If one cell is shaded 50%, the module output will decrease approximately 50%. It is interesting to note that if one entire row or column of cells is 50% shaded, the module efficiency will decrease 50%, due to the series wiring of all cells in a module.[1]

Shading obstructions can be defined as soft or hard sources. If a tree branch, roof vent, chimney, or other item is shading from a distance, the shadow is diffuse or dispersed. These soft sources significantly reduce the amount of light reaching the cell(s) of a module. Hard sources are defined as those that stop light from reaching the cell(s), such as a blanket, tree branch, or bird dropping sitting directly on top of the glass. If even one full cell is hard shaded, the voltage of that module will drop to half of its unshaded value in order to protect itself. If enough cells are hard shaded, the module will not convert any energy and will, in fact, become a tiny drain of energy on the entire system.

---

[1] Kyocera Modules, Solar Products Catalog, Kyocera, Inc.

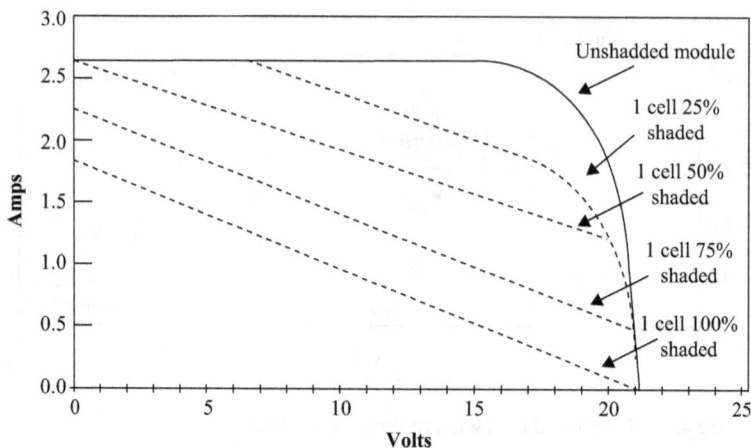

Figure 4.12. Shading effect on the PV module.

To mitigate the shading problem, most commercial PV modules come with a bypass diode built into the module output wiring. The diode serves to direct current around the module if the module in a string is shaded and not generating current. The diode will prevent the unshaded modules from feeding the shaded module and prevent excessive power loss.

4.4 PHOTOVOLTAIC ARRAYS

PV arrays consist of two or more modules wired in series or parallel to obtain a desired voltage, current, and power output. Arrays may consist of one or more strings of modules. Modules in a string are wired in series, and strings are connected in parallel to the inverter. If the array is con-nected to a single inverter, all modules in the array must be installed at the same tilt angle and orientation. Also, modules in the array should have identical electrical characteristics. Ideally, the modules should be manu-factured in the same lot and sorted by electrical test results. The reason for these installation criteria follows from the previous discussions about *I–V* curves, inverter operation, and MPPT.

When two modules are wired together in series, the combined output voltage is the sum of the individual module voltages, and the output current is only slightly more than the current of the module of lowest rating. Figure 4.13 shows two modules of different ratings that have been wired in series. Module A is rated 21 V and 5.4 A, and module B is rated 21 V and 2.8 A. These modules will generate about 113 W and 59 W, respectively. When

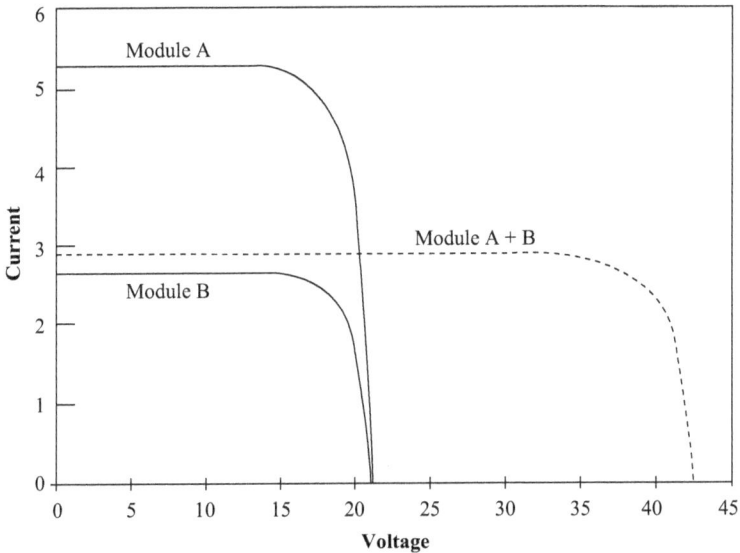

Figure 4.13. Series connected PV modules.

wired in series, the total maximum output is about 126 W. That is 46 W less or about 27% less than individual nameplate ratings. This example exaggerates the loss of potential array power if module electrical characteristics are not matched, but the principle holds true for all PV systems, unless each module is connected to individual inverters.

If we connect two dissimilar modules in parallel, output voltage will be the average of the rated voltage for each module, and power output will be less for the parallel array than individual arrays for each module. If module A is rated 25 V and 3 A, and module B is rated 21 V and 2.8 A, the combined parallel output will be about 133 W. If the modules are in separate arrays, they would produce 75 W and 59 W, respectively, which are about 1% more power. Therefore, voltage matching is not as critical as current matching in arrays, but still a source of power loss (Figure 4.14).

4.5 PHOTOVOLTAIC ARRAY DESIGN

Equipped with the information on PV cells and modules, let us design a PV array and interconnected system for a string inverter. The first piece of information needed is the energy requirement for the application—residential or small commercial. Let us assume that we are designing a roof-mounted PV system for a residential application that historically uses

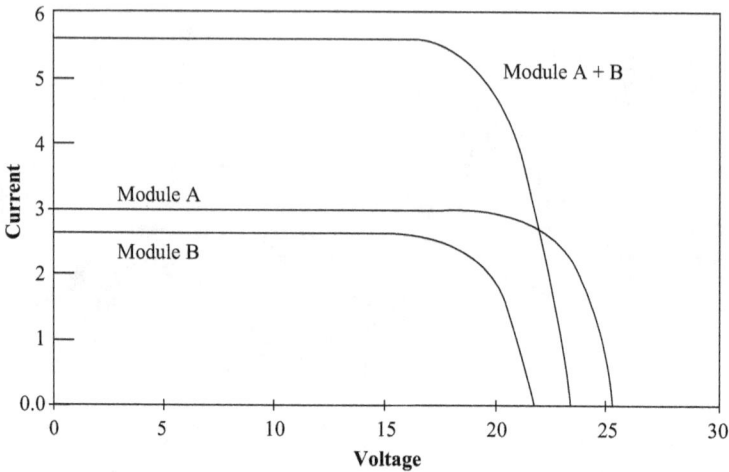

Figure 4.14. Parallel connected PV modules.

6,000 kWh/y, or about 500 kWh/month on average. We will assume that the premise does not have a demand tariff rate. Energy usage information may be obtained from the utility or customer billing records. Economically, it is best to design a system that will produce annual energy equal to usage, because any excess energy may be sold to the utility at a reduced rate (e.g., 4.6 vs. 9.8 cents/kWh). Also, many utilities have a renewable energy capacity limit of 120% of usage. A rule-of thumb-estimate (1 kW system ~ 1,500 kWh energy/y) tells us that a 4 kW PV system will produce approximately 6,000 kWh/y.

The next step is to complete a site analysis, which will determine if we have any shading problems and confirm energy output. We can perform the analysis manually or use one of the electronic instruments described in the previous chapter. Let us use the iPhone app and compare the results with those of the PVWatts™ program (http://www.nrel.gov/rredc/pvwatts). The roof pitch (rise/run) is 5/12 for an angle of 22.6° (θ = tan^{-1} [5/12]) and the orientation (azimuth) is due south (180°). The location as determined by the global positioning system is 40° north latitude and −105° west longitude at an altitude of 1,632 m. The nearest weather station for clear sky data is Boulder, Colorado.

The SolmetriciPV app gives us total available radiation of 48,956 kWh/y. AC energy is estimated at 5,991 kWh/y for a 4.2 kW system at a system derating factor of 84.6% (cabling, connectors, array soiling, etc.). The application report also provides an energy-to-power ratio of 1,426 kWh/kWp. (We originally estimated system size using the rule-of-thumb ratio of 1,500.)

The PV watts program for a 4.2-kW system estimated an AC output of 5,942 kWh/y, with a default derating value of 77%. Both methods provided a daily average radiation of 5.39 kWh/m^2/day. Data results from the two methods are similar, except that the iPV program uses actual equipment specifications to determine a derating factor (see Appendix A for example reports and a list of derating factors).

Now that we know how much energy is required, we can select equipment that will produce approximately this energy given the roof space available. We carefully measure the dimensions of the roof to determine how much area can be covered with PV panels, allowing for spacing at the edges and avoiding any vents or chimney obstructions on the roof. We also must conform to local building codes that may determine the distance from roof edges and how we can attach the rails to the roof to meet wind-loading criteria. The total available roof area for PV panels determines the type of modules we can purchase. A small area may require high-performance panels (>15 W/ft^2). Let us assume that we have limited space and opt for single-crystalline modules rated at 200 W and are 13 ft^2 in area each.

String layout design is sometimes a trial-and-error process. For a 4,000 W system, we need 20 200 W modules. A two-string system with 10 modules per string appears to be a good starting point. We check the open-circuit voltage specifications for the 200 W module, and learn that open circuit voltage (VOC) = 70 V. A series string of 10 modules will give us 700 Vdc per string. NEC requires all residential equipment to be rated at 600 V maximum. To meet this requirement, we can design 3 strings of 7 modules each, and VOC for each string will be about 490 V. Another option is to find a different PV module that has a lower open-circuit rating.

The NEC also requires that the maximum PV system voltage shall be calculated as the sum of the rated open-circuit voltages of the series-connected modules corrected for the lowest expected ambient temperatures [20]. Table 690.7 of the NEC gives us a maximum correction factor of 1.25 for an ambient temperature of −40°C. However, the NEC also states that if temperature coefficients are supplied with the module instructions, they shall be used. Let us select a module with a temperature coefficient (VOC/°C) of −0.172 V/°C (Kt). Checking with our local weather station, the lowest expected temperature for the site is −30°C. The revised VOC for this module is calculated in Equation 4.5:

$$\text{VOC (max)} = \Delta T \times Kt + \text{VOC (at STC)}$$
$$= [-30 - 25] \times [-0.172] + 70v = 79.46 \text{ V} \qquad (4.5)$$

The maximum VOC for a seven-module string is 556 V, which is below the 600 V maximum.

The final step in the trial-and-error process is finding an inverter that meets voltage and power calculations for the array. The PV module specifications provide a maximum power voltage (V_{mp} = 55.8 V), maximum power current (I_{mp} = 3.59 A), and rated power (P_{max} = 200 W). The array of three seven-module strings connected in parallel will generate 4,200 W, 10.8 A, and 390.6 V at MPP. The inverter must also be rated for the maximum short-circuit current of 13.8 A (3 × 4.6), and maximum open-circuit voltage of 556 V. Inverter manufacturers allow the inverter rating to be 10% less than the PV array power, because the modules will operate below STC conditions in the field. A 4,000 W inverter is generally acceptable, but the manufacturer specifications must be examined before making this assumption.

The PV system must be designed so the PV array voltages are within the MPPT range of the inverter. Calculate the minimum and maximum voltage of the array at high and low temperatures for the installation location. The inverter MPPT range must include the maximum and minimum voltages. A common design error is neglecting to account for voltage derating over time. PV modules will degrade about 0.5%–1% per year. After 1,015 y of operation, DC voltage may drop below the minimum MPPT range of the inverter, and the system will shut down. Many inverter manufactures have online programs that will match their inverters to most any PV module available in the market. They also provide guidelines on how to connect the module strings and arrays to maximize peak power tracking and meet NEC requirements.

The final design may not exactly match the original estimates for energy generation, but the original energy calculations were based on energy use that will vary. Energy output will also vary with weather patterns and cloud cover variability. The 4.2 kW system will probably exceed estimated production because the PVWatts program is conservative with efficiency derating factors.

To summarize the PV array design process:

1. Perform energy analysis using PVWatts at the installed location and historical usage data.
2. Select array size and PV modules.
3. Determine string size and voltages.
4. Select an inverter; check MPPT, maximum and minimum voltages.
5. Calculate wire size and voltage drops.
6. Apply derating values for wire ampacity.
7. Size protective devices (fuses and circuit breakers).

8. Check service panel ratings.
9. Evaluate supply-side or source-side connections.

4.6 WIRING METHODS

Wiring the PV modules and inverter is regulated by the NEC, and guided by good practice for system efficiency. Voltage drop (Vd) on the DC side and AC side of the inverter is calculated by the following formula:

$$Vd\,(\%) = 100\,(1/V_s) \times R(\Omega/\text{kft}) \times (2 \times d \times I/1{,}000)$$
$$= 0.2\,(I \times d/V_s) \times R(\Omega/\text{kft}) \tag{4.6}$$

where,
V_s is nominal voltage
$R\,(\Omega/k\text{ft})$ is the resistance of the wire per 1,000 ft
d is the length of wire
I is the current

Wire resistance can be found in Chapter 9, Table 8, of the NEC. AC voltage drop limits provided by the NEC are 5% for total branch and feeder circuits, and 3% for either branch or feeder circuits. A feeder connects the utility to the circuit breaker panel, and a branch circuit connects loads to the panel. A 2% voltage drop maximum is recommended as a good practice for the DC wiring between inverter and modules to minimize resistive heating loss in the wiring (Figure 4.15).

The NEC specifies the type of wire required, conduit sizing and type, and wiring methods. Article 690 of the NEC is specifically for PV solar electric systems. Article 210—Branch Circuits, Article 240—Overcurrent Protection, Article 250—Grounding, and Article 310—Conductors, are the most common references for PV installations. Type THHN THWN-2 moisture and oil-resistant thermoplastic insulation rated at 90°C is recommended for PV installations. Article 310 specifies wire ampacities and references derating factors for specific installations (e.g., conduit above a roof).

Wiring must be enclosed in metallic conduit if the path enters the structure according to NEC section 690.31 (E). The maximum circuit current is 156% of I_{sc} as determined by 690.8 (A) and (B). Normal circuits require 125% wire ratings over the continuous current ratings, but the code requires an additional 125% due to the possibility that PV systems may peak 25% higher than rated when partly cloudy conditions act as a magnifying glass and increase solar radiation over the design rating of

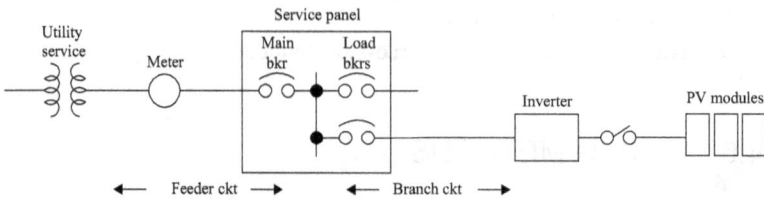

Figure 4.15. Utility feeder and branch circuits.

1,000 W/m². Continuous rating means greater than 3 h, and the NEC predicts that this condition may exist that long.

Total voltage drop in the DC circuit should be less than 2%. The DC circuit consists of wiring between the DC disconnect and the roof junction box, plus the interconnect wiring to and from and between modules. The first design step is to size the wire used between the inverter DC disconnect and roof junction box. Let us assume that this distance is 100 ft and ambient temperatures do not exceed 100°F (i.e., attic temperatures).

If temperatures exceed 30°C, and the wiring is in conduit, the current carrying capacity of the wire must be derated by a correction factor given in NEC Table 310.16. If attic temperature is anticipated to be 38°C, we must derate the conductor ampacity by 0.91 for conductors with insulation ratings of 90°C. If the conduit is above the roof at any point in the path to the junction box, a temperature correction factor must be added before using Table 310.16. Table 310.15 (B)(2)(C) provides the temperature adder for various heights above the roof. For example, if the conduit is 4 in above the roof, we must add 17°C to 38°C for an anticipated maximum temperature of 56°C. Now the correction factor from Table 310.16 is 0.71.

If more than three current-carrying conductors are installed in a conduit, NEC Table 310.15(B)(2)(a) provides an adjustment to the ampacity values provided in Tables 310.16 through 310.19. For example, if four to six current-carrying conductors are pulled in the conduit, maximum current is limited to 80% of the full conductor rating. Non-current-carrying conductors are ground wires that terminate at one point in the junction box and do not carry current under normal conditions. Non-current-carrying conductors are for the specific purpose of establishing ground potential for all-metallic equipment in the system. Current-carrying conductors carry the DC source current (to and) from the PV system to the inverter. Although the DC negative may be connected to the ground at some point in the system, the negative DC conductors must not be considered ground conductors.

After reviewing all the derating factors and adjustments, we determine that #10 copper THHN insulated wire has a derated ampacity of

29.1A. The maximum short-circuit current for the 4.2 kW array is 21.5 A, and the wire size is adequate to carry the current without overheating. It is a good practice to set up a table to identify all wire ampacity ratings, maximum current requirements, and derating factors, as shown in Table 4.3.

After determining the wire size, voltage drop can be calculated using the resistance values found in Chapter 9, Table 8, of the NEC. Insulated and stranded #10 copper wire resistance is 1.29 ohms per 1,000 ft (Ω/kft). The distance to the roof enclosure is 100 ft, $V_s = V_m = 55.8 \times 7 = 390.5$ V, and Vd, determined by Equation 4.3, is 0.91%. The voltage drop from the junction box to the modules must also be added. If the module loop is 50 ft ($d = 25$ ft) and #10 wire is used, voltage drop is 0.23%. The total voltage drop is 1.14%, within our 2% limit.

Other design considerations include sizing of ground conductors and ground-fault-protection equipment, which is determined in NEC Articles 690.45 and 250.122. For small systems, the ground conductor in conduit can be sized the same as the current-carrying conductors if protected by the same wireway or conduit. Wire size also depends upon fuse sizing (if used) and circuit breaker size. Ground-fault detection and interruption (GFDI) systems are usually built into the inverter. Inverters with GFDI protection must open the DC breaker upon faulty conditions. For exposed equipment ground wire, code requires #6 solid copper bare wire, which is more durable than stranded #10 wire. The DC circuit must include a 600 V rated disconnect and a fuse or circuit breaker sized greater than the maximum current of the array, but below the inverter-rated DC input.

The AC connection to the utility or point-of-utility connection (PUC) must include a circuit breaker or disconnect sized according to NEC Articles 690 and 240. The circuit breaker connected to the inverter must not be less than 125% times the inverter output rating. If the inverter AC output rating is 16.7 A, a 20 A circuit breaker is not quite adequate (125% × 16.7 = 20.8). The next size breaker must be used (30 A).

NEC Article 690.64 describes the AC panel requirements. The circuit breaker connected to the inverter is considered a "source" connection to the service panel. Therefore, the panel has two sources: the utility main

Table 4.3. Ampacity

I_{sc} A × I_{sc} A 1.56	I_{sc} A	Wire gauge #	30°C ampacity	Ambient tempature	Ambient derating	Conduit fill derating	Derated ampacity	I_{max}/ ampacity
13.8	21.5	10	40	38°C	0.91	0.8	29.1	0.74

breaker and the PV inverter connection. The sum of the two sources must not exceed the busbar rating of the service panel. For example, if the utility feeder breaker is rated at 100 A and the PV breaker is 30 A, the busbar rating must be at least 130 A. NEC 690.64 also states that the busbar rating may be exceeded by 20% if the source circuit breakers are located at the opposite ends of the service panel and labeled "do not relocate this breaker." Kirchhoff's current laws provide reason for this exception. Supply current will be tapped into loads along the panel, so the busbar will not carry the total supply current. As an example, if the main breaker connecting the utility is rated 100 A, and the PV inverter breaker is rated 20 A, the busbar rating may be 100 A only if the source breakers are located at opposite ends of the service panel and labeled properly.

Figure 4.16 summarizes the design in a modified one-line diagram, as required by the inspector and utility. The two-pole circuit breaker in the service panel that connects the inverter is rated 30 A, and the service panel busbar must be rated for the main breaker rating plus the 30 A PV breaker rating. A 150 A-rated service panel with a 100 A main breaker would be one alternative. A 200 A-rated service panel with a 200 A main breaker would also meet code if the inverter breaker is located at the bottom and labeled "do not relocate breaker."

NEC codes can be difficult to interpret, but must be followed to meet building codes. We do not have a simple code system that applies to every jurisdiction or anticipates every situation. NEC and local building codes are interpreted and applied by local jurisdictional building departments. When in doubt, consult local authorities before installing.

Grounding is a good example of local code interpretation. Some jurisdictions require a separate grounding conductor from the PV array to the building ground according to the 2008 NEC. Other jurisdictions allow the equipment-grounding conductor within the conduit from the array to the inverter, according to 2011 NEC. While it may appear to be good practice to ground the array separately for lightning protection, it may not be practical on all roof configurations. Some jurisdictions also require the electrical ground wire to be bonded to the main water piping and the foundation ground (if available). Again, it is best to consult local authorities before installing ground systems.

4.7 PV ARRAY DESIGN WITH MICROINVERTERS

Microinverters eliminate the DC array design described in the preceding section. The installer matches the PV module to the microinverter input specifications, and the high-voltage DC design is mitigated. However,

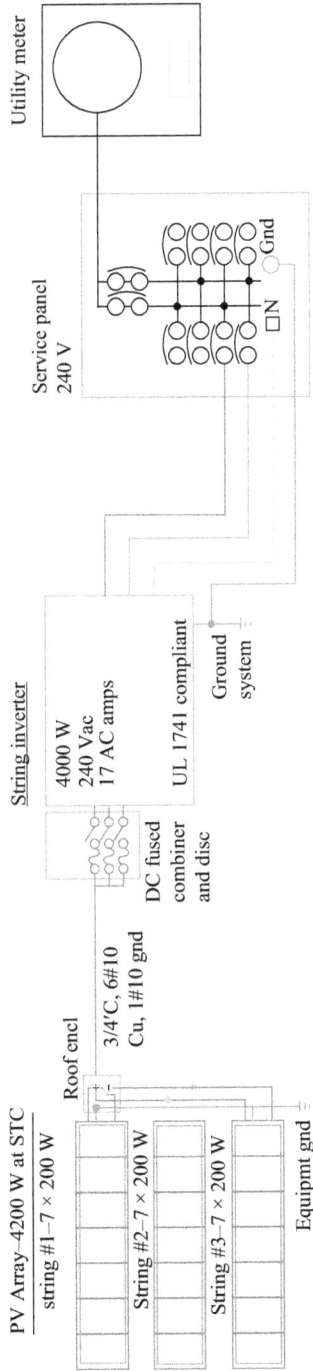

Figure 4.16. Line diagram.

now the installer must carefully design the AC connections to the electrical service panel (Figure 4.17).

Microinverter specifications must match individual PV modules. If the module and microinverter are purchased together as a package, the manufacturer has matched specifications. If purchased separately, the installer must ensure that the microinverter is designed to operate within the PV module voltage and current ranges. Sixty-cell PV modules typically produce 30–34 V at STC, and microinverters are designed for peak power tracking (MPP) of 25–40 V. At low temperatures, the 60-cell module may produce 44 V, and the microinverter must be designed for maximum DC input not less than 44 V. A 72-cell or larger PV module will produce voltages that do not meet the MPP tracking range or the maximum input voltage specifications of a microinverter. At high temperatures, PV module voltage may drop below the tracking range of the microinverter, and energy will be lost as the microinverter shuts down. As PV modules age, voltage degrades, and the module voltage may drop below the minimum tracking voltage—and, again, the system will shut down and energy will be lost.

The decision to install a central inverter or microinverters is dependent upon the PV array location, orientation, and shading. Central inverters are a good choice if the PV array is homogeneous. That means all modules in the array have identical orientation and tilt angle, and that shading is not an issue. If the array installation includes many different roof orientations or tilt angles, microinverters will maximize energy production because each PV module operates at its own maximum power tracking point. Also, if some modules are shaded during the solar day while other modules are unshaded, microinverters will save energy that will be lost with a central inverter. Partial shading of only a few modules can cause significant energy loss for an entire string, as described earlier in this chapter. Of course, equipment costs and installation costs should also be considered in the decision.

EXAMPLE PROBLEM

Example Problem 4.1

Design a PV system with microinverters and 240 W polycrystalline modules for a total DC power rating of around 3,500 W. Specifications for the PV module and microinverter are given in Tables P4.1.1 and P4.1.2:

The power rating of the PV module (240 W) is within the power range of the microinverter, the maximum peak power point (30.4 V) also

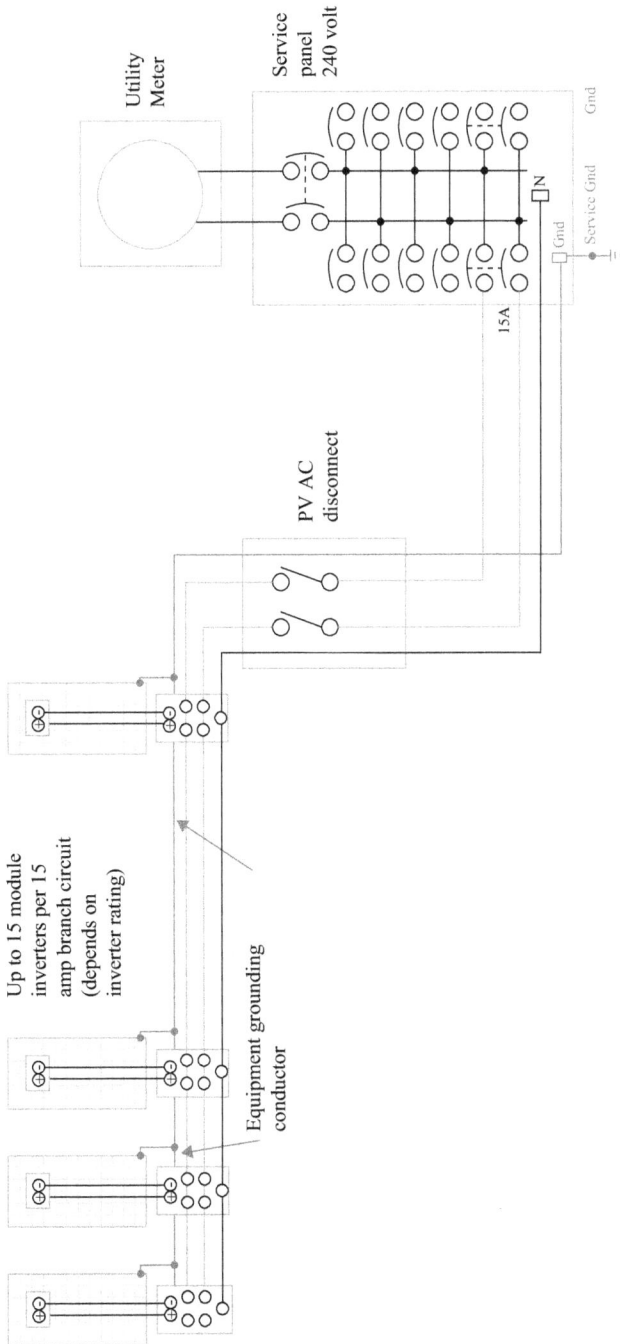

Figure 4.17. Microinverter schematic.

is within the inverter peak power tracking voltage range (22–36 V). The maximum module open-circuit voltage is determined by the low ambient temperature expected at the installed location. For –30°C, the maximum open-circuit voltage is 37.3 + 37.3 × (–0.0032) × (–30 – 25) = 43.9 V. The maximum module voltage is less than the maximum DC inverter voltage (45 V).

We also must look at the peak power tracking voltages. Applying the voltage coefficient at –30°C to the MPP voltage (30.4 V), the module maximum tracking voltage (V_{mpp}) is 35.75 V, which is within the inverter range (22–36 V). We also should check the minimum power point voltage of the module. For an expected high temperature of 40°C, the module minimum tracking voltage is 28.9 V, well above the inverter minimum tracking range. If we consider module degradation of 20% in 20 y, the minimum module voltage drops to 23.1 V. We are still within the peak power tracking range of the microinverter.

Now that we are comfortable with the voltage and power matching of the module and inverter, we can proceed to the AC design. To obtain about 3,500 W of DC power, we need about 15 modules and inverters (3,500/240). Fifteen modules will generate 3,600 W of DC power at STC, and the total

Table P4.1.1 Microinverter specifications

Recommended input power	190–260 W
Maximum DC voltage	45 V
Peak power tracking voltage	22–36 V
Maximum input current	10.5 A
Max output (AC) power	215 W
Nominal output current	0.90 A
Nominal output voltage	240 V AC

Table P4.1.2 PV module specifications

V_{mpp} at STC	30.4 V
I_{mpp} at STC	7.9 A
VOC at STC	37.3 V
Nominal power Wp	240 W
Power coefficient	–0.44%/°C
Voltage coefficient	–0.32%/°C

AC current for the parallel inverters is 13.5 A (15 × 0.90). Following NEC requirements, we must size the branch circuit breaker to 125% of the AC inverter rating, or 16.9 A. The next larger size circuit breaker is 20 A.

The AC conductor for the wiring from the PV array to the inverter is based on the total output current of 13.5 A and the protective circuit breaker. One hundred twenty-five percent times the rated current is 16.9 A, and #14 copper THWN-2 appears to be adequate. However, NEC Article 240 requires that #12 copper wire minimum is to be used with a 20 A circuit breaker. Also, correction factors for ambient temperature and number of conductors must be applied (see NEC Article 310). We also should check voltage drop if the PV array is located a distance more than 100 ft from the electrical service panel. Maximum voltage drop for the feeder circuit should be less than 2% (see discussion on wiring methods in Section 4.6).

Finally, the microinverter is compatible with the 240 W PV polycrystalline module selected, and we can install 15 modules on one 20 A feeder circuit. The system is rated at 3,600 W at STC.

4.8 OPTIMIZERS

Another option for inverter design is the power optimizer. This design uses a string or central inverter with DC–DC voltage converters on each module in the array. The module voltage converter is called a power optimizer, and has the advantage over normal string design by allowing each module to operate at its own MPP. The string voltage is controlled by the central inverter, and is usually fixed at around 500 V DC. Since each module operates at its own MPP, this design mitigates shading effects of individual modules. If one module is shaded or otherwise underperforming, the remaining PV modules in the string continue to operate at maximum power. Optimizers also allow more modules per string—up to 25, depending upon the power rating of the inverter. The minimum number of modules per string is eight.

Another advantage of the optimizer design is that strings do not need to have an equal number of modules. Optimizers allow each module to operate at MPP regardless of the string voltage at the inverter. Most conventional central inverters have one MPPT function for the entire array, and parallel strings must be equal to ensure that all modules are operating close to their MPP. Optimizers let each module operate at individual MPPs, and also adjust individual output so that the inverter voltage is constant. Another advantage of the optimizer design is flexibility of module orientation. Since each module operates independently, modules do not need to be installed at the same pitch and orientation. Modules may be installed on different roof areas of varying pitch and orientation and still remain on the same string. Conventional string design required all modules in a string to be installed on a homogeneous roof area.

A disadvantage of optimizer design is the additional equipment cost and additional labor to install optimizers under each module. Unlike microinverters, a special central inverter must be installed with optimizers. As with microinverters, the added equipment can also mean more equipment that may fail. It can be difficult to replace or repair since optimizers and microinverters are located under the array on the roof.

Figure 4.18 is a wiring schematic for the optimizer design. String A and String B have different numbers of modules (12 and 11, resp.). The two strings are paralleled at the inverter. Note that this system could also be designed with one string of 23 modules!

The drawing shows how PV cables connect modules in each string in series. An equipment grounding conductor ensures that all modules, rails, optimizers, and other PV hardware are connected to the electrical building ground.

PROBLEMS

Problem 4.1

Redesign a nominal 4 kW PV array using Kyocera polycrystalline 200 W modules and a string inverter. What is the new maximum open-circuit voltage at −30°C? (See the specification sheet in Appendix B.) How many modules would you connect in a series string? (Adjust the array size to match the string design with the new modules.)

Problem 4.2

Estimate the AC power output for the new array in Problem 4.1. Use the PVWatts program with the same location information.

Problem 4.3

Run the PVWatts program with data from Problem 4.2 for a solar orientation of 150°, 210°, 120°, and 90°. What is the percentage reduction in estimated AC power output for these orientations?

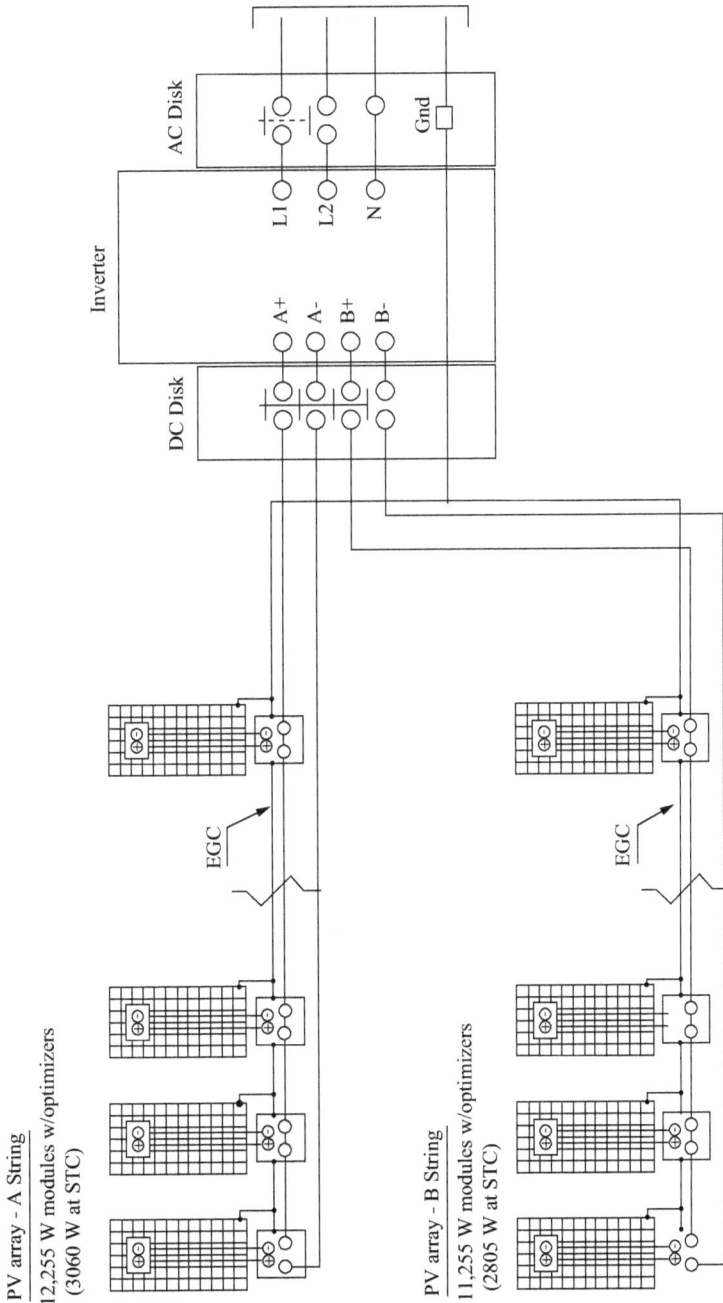

PV array - A String
12,255 W modules w/optimizers
(3060 W at STC)

PV array - B String
11,255 W modules w/optimizers
(2805 W at STC)

Figure 4.18. PV system design with optimizers.

Problem 4.4

Run the PVWatts program with data from Problem 4.2 for a tilt angle of 40° (default). Repeat PVWatts for tilt angles of 30°, 20°, 10°, and 0°. What is the percentage reduction (from the default) in estimated AC power output for each tilt angle?

Problem 4.5

Calculate the voltage drop on the DC circuit if the distance from the inverter to the roof enclosure is 50 ft, array loop is 40 ft, and you are using #10 THHN copper wire in the conduit run and #10 USE in the array loop. Note: d in the Vd formula is the total distance from the inverter to the middle of the array loop, and $2d$ is the complete run and loop distance to and from the inverter.

Problem 4.6

Use the inverter and module sizing program provided by Fronius or SMA-America on their websites to design an 8 kW array using polycrystalline modules (200 to 230 W). Assume 30°C minimum, 40°C maximum ambient, and sufficient roof spacing for 40 modules. How many modules are in a string?

Problem 4.7

Determine the AC-maximum-rated current output of a 4 kW inverter. What size wire and circuit breaker is required? (See NEC Article 240.)

Problem 4.8

Design a microinverter system using 230 W monocrystalline modules and 215 W inverters. What is the maximum number of modules and inverters possible on a 20 A branch circuit?

CHAPTER 5

INTERCONNECTED WIND ENERGY SYSTEMS

As mentioned in Chapter 2, wind energy is the fastest growing renewable energy source in the United States. The growth rate for wind is about 30% per year, compared to 21.6% for photovoltaics (PV), and 1% for coal [21]. Large-scale wind farms have taken over the industry that once consisted of small, stand-alone wind generators that charged batteries for remote homes. The 10 to 20 kW wind machines of those times are now dwarfed by 1 to 3 MW wind turbines on 300 ft towers. Reliability and electrical performance have improved to the point where the annual production of wind farms can be more accurately predicted and managed by the interconnected utilities. This chapter will describe the wind resource in this country and the basics of aerodynamics, generator operation, and balance-of-system equipment associated with interconnected wind turbines.

5.1 THE WIND RESOURCE

Wind energy is an indirect form of solar energy. Thermal gradients caused by solar radiation cause air mass movements. Rising air at the equator is replaced by air movement from the north and south latitudes. Rotation of the Earth creates trade winds east to west in some parts of the world, and west to east in other regions. Diurnal winds are created by land and sea thermal gradients. Weather systems create synoptic winds that are less predictable. Turbulence is a characteristic of synoptic winds that is not a favorable component of wind when extracting energy. Turbulence disrupts the Bernoulli effect and causes airfoils to lose lift and efficiency. The best wind resources are areas of consistent and uniform wind characteristics.

Table 5.1. Wind energy classification

	Hub height					
	10 m		30 m		50 m	
Wind class	Power density (W/m²)	Wind speed (m/s)	Power density (W/m²)	Wind speed (m/s)	Power density (W/m²)	Wind speed (m/s)
1	100	4.4	160	5.1	200	5.6
2	150	5.1	240	5.9	300	6.4
3	200	5.6	320	6.5	400	7.0
4	250	6.0	400	7.0	500	7.5
5	300	6.4	480	7.4	600	8.0
6	400	7.0	640	8.2	800	8.8
7	1,000	9.4	1,600	11.0	2,000	11.9

The wind resource for the United States has been mapped in terms of wind speed and wind power density in watts per square meter. The U.S. Department of Energy and National Renewable Research Laboratory have developed a wind speed map from wind speed records (mostly airport data) throughout the country. These maps are available on the Internet at www.nrel.gov/wind/resource_assessment. Seven wind classes categorize average wind speed and power density.

Table 5.1 shows the wind classifications, wind power density, and average wind speed for three hub heights. Hub height is the distance above the ground where an anemometer is installed to measure wind speed.

The U.S. wind resource is large enough to produce more than 3.4 trillion kWh of energy each year [22]. Almost 90% of the available wind resource lies in the Great Plains area. North Dakota has about 36% of the Class 4 wind resource area that is available for wind generation development. However, much of this resource is located far from the urban areas where the electricity is needed. Transmission lines must be added to make wind energy feasible in these areas. Other constraints to wind development include public acceptance, institutional constraints, access, and technological problems.

Kinetic energy (KE) of an air mass (m_a) moving at speed V (m/s) is given by the following equation:

$$KE = \frac{1}{2} m_a V^2 \text{ J} \qquad (5.1)$$

Power is the flow rate of kinetic energy per second in watts:

$$P = \tfrac{1}{2} \text{ (mass flow per second) } V^2 \qquad (5.2)$$

From these relations, we may determine the power from a volumetric flow of air mass given by Equation 5.3:

$$P = \tfrac{1}{2} (\rho A V) V^2 = \tfrac{1}{2} \rho A V^3 \qquad (5.3)$$

where P = mechanical power in watts (kg m²/s³)
ρ = air density (kg/m³)
A = area swept by the rotor blades (m²)
V = velocity of air (m/s)
If specific sites are being evaluated, it follows that the wind resource should be expressed in terms of the wind power per square meter (W/m²). Equation 5.4 represents the *specific windpower density (SP)* for the wind resource:

$$SP = \tfrac{1}{2} \rho V^3 \qquad (5.4)$$

Obviously, a wind generator cannot extract all the energy from the wind, because that would give claim to an efficiency of 100% and defy a number of physical laws. A number of books are available on the subject of aerodynamic efficiency, which we will not get into in this text. The efficiency of any rotor is customarily expressed as a coefficient and included in the power equation for *net power extracted from wind turbine blades*.

$$P = \tfrac{1}{2} \rho A V^3 C_p \qquad (5.5)$$

where C_p is called the *power coefficient* of the rotor, or rotor efficiency. It varies for different types of blades, rotor design, wind speed, and rotor speed. It has a maximum theoretical value of 0.59, but in practical designs, it remains in the range of 0.4 to 0.5 for modern two-blade turbines. Higher-density rotors (more than two blades) usually have a C_p between 0.2 and 0.4.

Wind power varies linearly with the *air density (ρ)* as in Equation 5.4. Air density is dependent on pressure and temperature according to the gas law:

$$\rho = p/RT \qquad (5.6)$$

where p = air pressure (lb. per square inch or psi)

T = temperature (absolute °K)

R = gas constant

The air density at *sea level* (ρ_o) at 1 atmosphere pressure (14.7 psi) and 60°F is 1.225 kg/m³. Air density at other elevations (Hm) in meters is corrected by the formula:

$$\rho = \rho_o - (1.194 \times 10^{-4}) \times Hm \text{ kg/m}^3 \qquad (5.7)$$

At an elevation of 2,000 m, air density is 0.986 kg/m³, which relates to a 20% difference in the power density calculation. We can roughly predict a 10% loss in power density for each 1,000 m of elevation gain.

The swept area of a rotor is the area of a circle in which the rotor rotates. Output power of a wind turbine varies linearly with the swept area. For a *horizontal axis machine*, the swept area (A) calculation is simply the area defined by the rotor diameter:

$$A = (\pi/4)\, D^2 \qquad (5.8)$$

where D is the rotor diameter.

If the wind turbine is a vertical-axis design (i.e., Darrieus), the area is $(2/3) \times$ (width at center) $\times$ (height).

The wind turbine intercepts the wind energy flowing through the entire swept area regardless of the number of blades. *Solidity* is defined as the ratio of the area of the blades to the swept area. A modern two-bladed wind turbine has a solidity ratio of 5% to 10%. As we will discuss later, the two-bladed machine is more efficient at high winds and less costly due to the lower blade material. The three-blade design boasts a 5% power performance increase, improved balance, and smoother operation, as compared to the two-blade design at the same angular speed. However, the third blade increases cost, weight, and installation work.

Wind speed measurements and methods of evaluation need further study because average wind speed is the most critical data needed to estimate power production for a particular site. Wind is highly variable by the minute, hour, day, season, and even by the year. Weather patterns generally repeat over a period of 1 y, and a 1 y site evaluation for average wind speed is a minimum. Ten-year evaluations are preferred, and give more confidence toward future energy potential. However, most projects cannot be put on hold for long wind evaluations, and 1 y wind speed data are often used to compare potential sites. This process is known as the *measure, correlate, and predict technique.*

Wind speed variations are best represented by a *probability distribution function*. The Weibull probability distribution is usually used to predict variability in wind speed. Weibull uses a shape parameter (k) and scale parameter (c) as given by the following expression:

$$h(v) = [k/c][v/c]^{(k-1)}e^{-(x)} \text{ for } 0 < v < \infty \qquad (5.9)$$

where $x = (v/c)^k$.

By definition of the probability function, the summation or integral of all wind speeds between 0 and ∞ must be equal to 1:

$$\int_0^\infty h \, dv = 1 \qquad (5.10)$$

The *shape parameter* (k) defines the wind characteristics of a potential site. If the site has many windless days, the Weibull distribution curve with $k = 1$ best represents the wind speed data. Most wind distributions take the form of the curve with $k = 2$, with a small number of windless days and a diminishing number of high-wind days. This special distribution curve ($k = 2$) is called a Rayleigh distribution, and is used for most wind site analyses. The Weibull distribution curve with $k = 3$ is closer to a bell-shaped curve with an equal number of high-wind and low-wind days. The bell-shaped curve ($k = 3$) is sometimes called a *Gaussian distribution* (Figure 5.1).

The *scale parameter* (c) shifts the wind speeds to a higher speed scale (the hump moves to the right for a larger c value). Usual values of c are 10–20 mph. Figure 5.2 shows the effect of increasing c from 8 to 12 mph, while the shape factor k is held constant at 3.

Most site data are reported in terms of mean wind speed, and are used to evaluate potential wind sites. If a more accurate prediction of power density is needed, *root mean cube* (rmc) wind speed may be used. V_{rmc} is

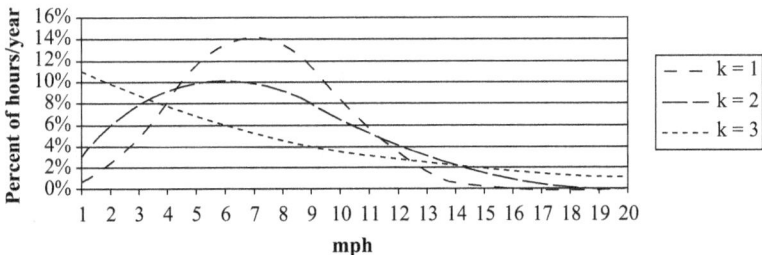

Figure 5.1. Weibull distribution and shape parameter (k).

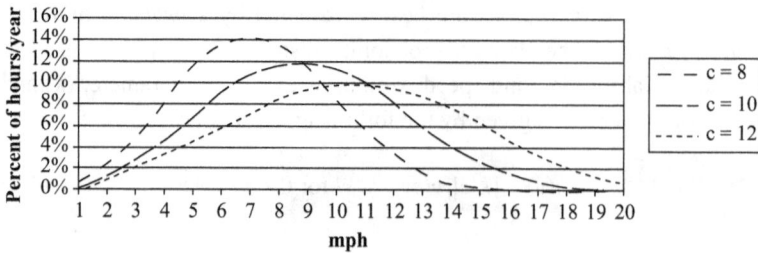

Figure 5.2. Weibull distribution and scale parameter (c).

calculated by taking the cube root of the integral of $hv^3\, dv$. If the data are from a 1 y evaluation with hourly average wind speed values, the integral is divided by 8,760 h.

$$V_{rmc} = [(1/8,760)\int hv^3 dv]^{1/3} \qquad (5.11)$$

Using V_{rmc} in the power formula (Equation 5.4) gives us the annual power generation in W/m². The root mean cubed wind speed calculation provides another look at potential wind sites when the average or mean wind speed provides insufficient information to decide if the potential site is feasible.

Actual data from a wind site in North Dakota [23] are shown in Figure 5.3. Instrumentation captured the time the wind speed was within twenty 1 m/s intervals (called bins). Figures 5.3 and 5.4 show the corresponding frequency of wind speed in each bin (1 m/s = 2.24 mph). The duration of this test was 20 days of very consistent high wind and good variability of speeds, which provided accurate performance data. The data were used to create a power curve for a 10 kW wind generator installed at the site. Test data were obtained using American Wind Energy Association performance standards that require a minimum duration in each wind speed bin. The data verify the Rayleigh distribution probability curve with a shape parameter of about 2.5 and a scale parameter of about 20 mph. The data are not representative of the yearly wind speed distribution due to the short period of test data.

Mode speed is defined as the speed corresponding to the hump in the distribution function (or, in this case, the actual data). The mode speed in Figure 5.3 is about 5 m/s or 11 mph. This is the speed of the wind most of the time. Figure 5.4 uses the same data as Figure 5.3, modified as a smoothed curve.

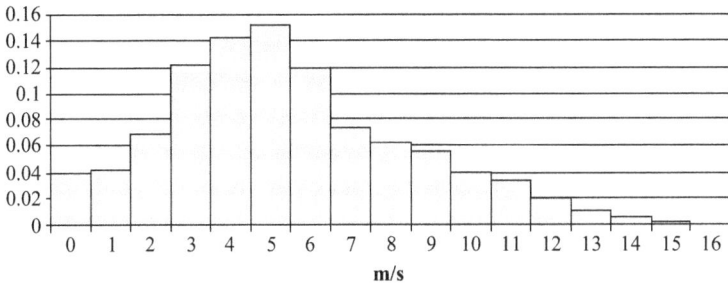

Figure 5.3. Frequency distribution of wind speed.

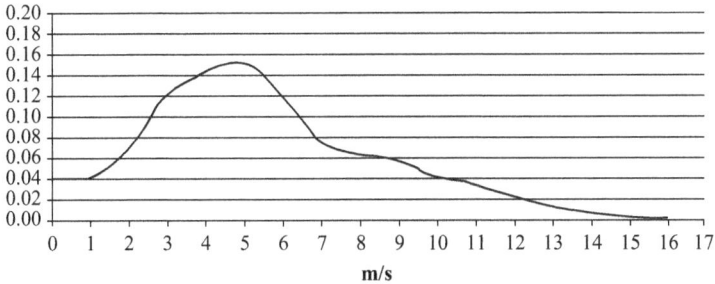

Figure 5.4. Smoothed frequency distribution of wind speed.

Mean speed is defined as the total area under the distribution curve integrated from 0 to ∞ and divided by the total number of hours in the period. If the period is 1 y, the area would be divided by the number of hours in a year (8,760).

$$V_{mean} = (1/8,760) \int_0^\infty hv \, dv \tag{5.12}$$

A thorough study of the probability of wind density distributions can involve some sophisticated mathematical approaches and three-dimensional analyses. We can avoid the math by making a few basic approximations that hold true for most wind distributions. The mean speed can be approximated by

$$V_{mean} = 0.90 \, c \tag{5.13}$$

where c is the scale parameter described earlier for the Weibull probability distribution.

This relationship may be used for the Rayleigh distribution with a shape parameter of $k = 2$ with reasonable accuracy.

Mean wind speed is usually obtained using digital processing techniques. Wind speed is sampled by a data collection system every few seconds. This sample is averaged into minute averages. Minute averages are then averaged into hourly, daily, monthly, and yearly averages. The averaging algorithm is given in Equation 5.14, where n is the number of samples.

$$V_{ave} = (1/n)\sum_{i=1}^{n} V_i \qquad (5.14)$$

If the root mean cubed velocity (V_{rmc}) is preferred for analysis, the following equations are used:

$$V_{rmc} = \left[(1/n)\sum_{i=1}^{n} V_i^3 \right]^{1/3} \qquad (5.15)$$

and

$$P_{rmc} = (1/2n)\sum_{i=1}^{n} \rho_i V_i^3 \qquad (5.16)$$

where ρ_i = air density (kg/m³)
v_i = wind speed (m/s)
n is the number of samples

The hub height of the anemometer used in making wind speed measurements must be corrected to the actual hub height of the proposed wind turbine. Many weather stations are located near airports where there are low-wind conditions, and the anemometer is installed 10 or 20 ft above the ground. Wind shear causes measurements to be lower near ground level. The following relationship corrects measurements at a lower level (h_1) to higher heights (h_2):

$$V_2 = V_1 (h_2/h_1)^\alpha \qquad (5.17)$$

where α is the ground surface coefficient
$\alpha = 0.1$ is smooth ground or a body of water
$\alpha = 0.4$ is in a city or around tall obstructions

Wind speed measurements are also more reliable if using ultrasonic anemometers that are not affected by cold, ice, rain, or bearing wear.

5.2   WIND POWER SYSTEM COMPONENTS

All wind systems have the following basic components: tower structure, rotor, nacelle, generator, speed control system, and yaw control. Larger wind generators also have a gearbox, braking system, and power electronics. This section will describe these components and their general design and operation.

The wind tower supports the rotor, generator, nacelle, and other components in the nacelle. Smaller wind turbines may use lattice or tubular towers, and large machines generally have concrete cylindrical towers. The height of the tower is usually equal to the rotor diameter for large machines to ensure that the blades are above any ground-induced turbulence or shear. Smaller wind turbines may have towers several times the rotor diameter in height to place the rotor in higher wind speeds and eliminate the effects of ground shear and turbulence. As described in the last section, wind speeds increase with height by a factor dependent on ground surface smoothness. Tilt towers, with hinges at the base and a gin pole leverage arm, are sometimes used to install smaller wind turbines. The tilt tower is also useful for maintenance and repair. Large machines are installed at the top of the tower with large cranes. With the exception of blade replacement, maintenance of the equipment in the nacelle is accessed with an elevator or climbing ladder inside the tower.

A primary concern in the design of wind turbine towers is structural dynamics. Wind speed fluctuations and rotor dynamics cause vibration at frequencies that may result in tower stress and fatigue. Resonant frequencies among the rotor, blades, and nacelle must be avoided as much as possible. Test facilities measure actual stress and strain caused from loads applied to the rotor and blades, and determine the critical frequencies of each component. Modal analysis uses complex matrix equations to identify these frequencies, and the results determine the design of the rotor, blades, and tower. Some of the analysis can be done in the lab (large test facility), but much of the data are obtained from measurements on machines operating on site. Resonant frequencies also may cause objectionable noise when wind speeds change suddenly.

The weight of the rotor and nacelle has a significant impact on the cost and design of the tower. This weight is called the top head mass (THM), and is minimized using as much lightweight material in the rotor and nacelle as possible without compromising strength. The largest wind turbines (5 MW) made today have a THM of 350 tons. The THM of the smaller Vestas wind turbine (3 MW) is 103 tons. The gearbox, electric generator, drive shafts, and rotor hub account for most of the THM, but steel must be used to withstand the stress on these components.

The nacelle houses the gearbox, generator, high-speed and low-speed shafts, brakes, and control equipment. A nacelle for large turbines also provides room for maintenance workers to access the components in a safe environment. It must be lightweight and strong to minimize THM and protect equipment from the environment.

The gearbox increases the rotational speed of the rotor (30 to 60 rpm) to the synchronous speed of the generator (1,200 to 1,800 rpm). As one can guess, the torque on the gearbox is very large, and cooling systems are usually required. The gearbox also contributes significantly to the THM. The gear ratio is usually fixed since shifting gears while a wind turbine operating under large torque is all but impossible. Direct-drive generators are used with small machines, and variable speed generators are also being designed to reduce or eliminate the need for a gearbox.

Modern wind turbines have two or three blades, which are carefully designed aerodynamic airfoils. Gone are the old, water-pumping flat or cupped blades that operate on the principle of drag. Blades apply the Bernoulli principle with a relatively flat bottom surface and longer upper-side surface. Lift is created on the airfoil by pressure difference in wind flow over the blade. Lift is increased by increasing the angle of attack slightly with pitch controls. The *angle of attack* is the angle between the wind direction and the chord of the airfoil. A drag force is also created, which impedes the lift force. The lift-to-drag ratio varies along the length of the blade to maximize power output at various wind and rotor speeds. Pitch control rotates the blade as the rotor speeds up to maximize lift and power to the shaft. Pitch control also is used to reduce lift and power as the turbine reaches maximum-rated power. When wind speeds exceed rated values, the pitch is changed to stall the blade and act as a brake.

In addition to pitch control for rotor speed and power, other control systems are needed to maximize power output, prevent overspeed, and prevent exceeding rated generator power. Yaw control continuously keeps the rotor facing directly into the wind. Many types of systems have been developed to control yaw, including the tail vane used by small wind turbines. The tail vane also acts as a brake when high winds occur by rotating the rotor (axis) perpendicular to the wind direction. Downwind machines use the nacelle as a wind vane. Larger wind turbines use a combination of power monitoring and wind direction detectors to control yaw. Yaw changes must be dampened to prevent noise generated by rapid yaw movement in high winds. Tilt or teeter is the angle of the rotor shaft to horizontal. Teeter may be increased to several degrees to reduce noise caused by tower interference and prevent flexible blades from contacting the tower.

Wind turbine blades must withstand extreme mechanical stress, vibration, and fatigue. Centrifugal forces, combined with stress and strain from wind forces, may cause premature failure if there is any flaw in the fabrication process. Recent improvements in rotor blade technology allow construction of durable blades over 50 m in length and rotor diameters over 100 m. The emergence of carbon and glass–fiber–epoxy composites is one of the primary advances in this technology.

5.3   WIND TURBINE RATING

Although there is no universal standard for rating performance of a wind turbine generator (WTG), there are a few generally accepted methods that manufacturers use in their specifications. The problem with rating wind generators arises because the output of a machine depends on the square of the rotor diameter and the cube of the wind speed. A rotor of a given diameter will generate different power at different wind speeds. The question is at what wind speed should a WTG be rated? Some manufacturers use a higher wind speed than others to show greater output for the same design. Rotor diameter rating also is not a good standard, because some wind turbines are designed to operate more efficiently at low wind speeds, while other machines are designed to withstand higher winds at the expense of low-speed efficiency. The best comparison of performance is a power curve that shows efficiency at a wide range of wind speeds, but these tests are difficult to control and expensive to perform on each machine design.

A comparative index known as the *specific rated capacity (SRC)* is often used as a reference in WTG design. It measures the capacity per square meter of the blade swept area and defined in units of kW/m^2:

$$SRC = \text{Generator electrical capacity/rotor swept area} \qquad (5.18)$$

For example, a 300/30 (kW/m) wind machine has a 300 kW electrical generator and 30 m diameter blade, and will have an SRC of

$$SRC = 300 \text{ kW}/(\pi \times 15^2) = 0.42 \text{ kW/m}^2 \qquad (5.19)$$

SRC ranges from 0.2 kW/m² for small machines to 0.5 kW/m² for large rotor diameter machines. SRC is a useful factor for the system integrator who needs to match the rated output of the generator to the interconnected-power equipment, including transformers, cables, and sub-station equipment. It also provides a measure of the blade design that will capture the available wind energy for the specific site.

The power curve for a wind generator shows power output for wind speed range from cut-in to cut-out. *Cut-in wind speed* is the minimum wind speed when the generator begins to produce power, and *cut-out wind speed* is that at which the WTG begins shutdown for protection from overspeed damage. The power curve of Figure 5.5 is from a 10 kW wind turbine. Full rated output is not reached until the wind speed reaches 15 m/s. The rotor diameter is 10.27 m, and the generator is rated 20 kVA for an SRC of 0.48.

A torque curve for the 10 kW wind generator is provided in Figure 5.6. Torque ($T = P/\omega$) in foot pounds per second is plotted versus angular velocity. The generator is rated at 10 kW at 25 mph wind speed (11 m/s) and 195 rpm. Torque levels at rated output and drops off sharply at cutout (not shown). The torque curve is typical of a variable speed WTG, since the generator is an alternator that produces power from 450 to 1,170 rpm. The alternator power output is rectified to direct current (DC) and con-nected to the utility by an inverter. The alternating current (AC)–DC–AC design allows the WTG to operate variable speed and capture energy over a wider range of wind speed and rotor speed.

Power and torque curves show that it is very beneficial to match the angular velocity of the rotor to the wind speed. The power coefficient (C_p) is higher when the angle of attack maximizes lift versus drag. This happens when the effective wind direction is at the ideal angle of attack.

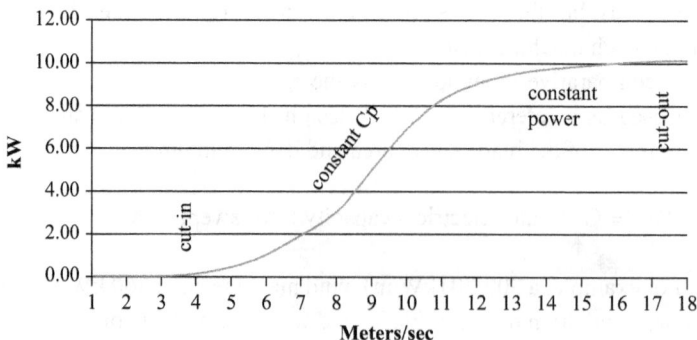

Figure 5.5. Wind turbine power curve.

Effective wind direction is determined by the orthogonal wind speed and the blade speed. It follows that the blade speed and wind speed have optimum values for maximum efficiency. This concept is embodied in a ratio called the *tip speed ratio* (*TSR*).

$$TSR = \omega \, r/v \qquad (5.20)$$

where v = wind velocity (m/s)
r = blade radius (m)
ω = angular velocity (radians/s)
The range for TSR is 1 to 6, depending on the blade design. Figure 5.7 shows several types of blade designs and the coefficient of power versus the range of TSR for several designs. The ideal rotor design operates at high efficiency over a wide range of TSR. This means that the rotor/blades perform efficiently over a wide range of wind speeds and angular velocities of the rotor. Actual rotor designs operate efficiently only for a narrow range of TSR—the wind speed and angular velocity must be within limits.

The modern two-blade design has the highest coefficient of power (i.e., efficiency) at a TSR range of about 4.0 to 7.0. The multiblade rotor, which is of the "water-pumper" farm-style design, has the lowest efficiency and operates at a very small TSR range, which means that angular velocity and wind speed must be relatively constant (*TSR* = 1) to operate efficiently. The multiblade and Savonius rotors rely on drag instead of lift. At low wind speeds, the drag component of wind provides the energy for this type of blade. At high wind speeds, efficiency drops off sharply since drag type machines do not operate at high shaft speeds. While the wind efficiency and energy are low for multiblade machines, an advantage is overspeed protection at high winds. Low efficiency at high winds act as a brake to slow the rotor.

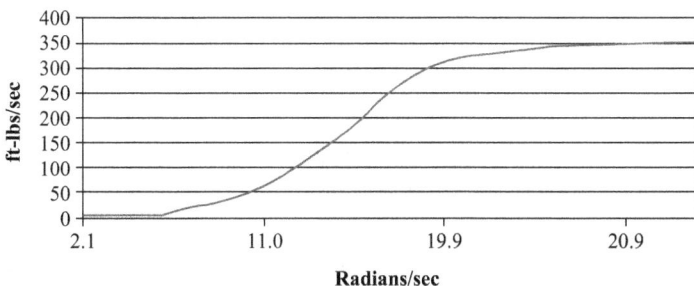

Figure 5.6. Wind turbine torque curve.

Typical wind machine blade performance

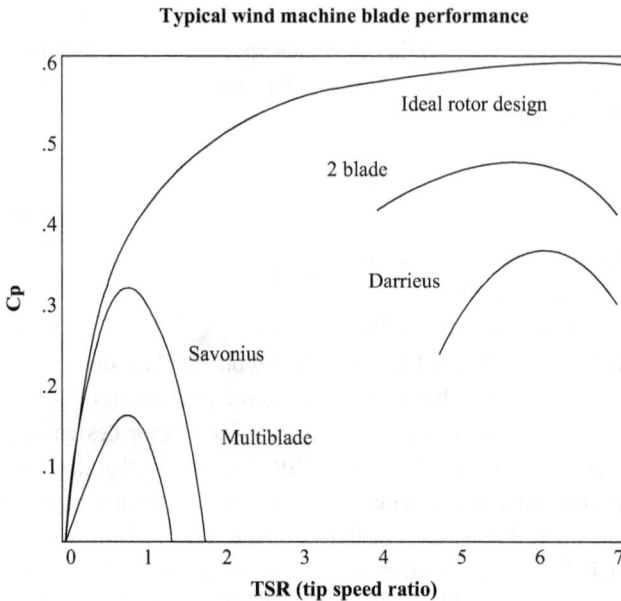

Figure 5.7. Tip speed ratio and power coefficient.

The Darrieus (vertical-axis) design operates well at high TSR numbers, but efficiency is very low in the low TSR range. At low wind speeds, a very small amount of lift is generated by the narrow blades, and a Darrieus rotor requires motor startup. As the rotor and blade speed approach wind speed, significant lift force is generated twice per revolution to maintain operation and generate significant power.

A three-bladed rotor design is the preferred option for some European manufacturers, as the added weight and cost of the third blade are not as significant to the additional 5% performance gain. A three-blade rotor has the advantage of smoother operation, improved balance, and lower harmonic and vibration issues. The TSR curve for the three-blade machine is slightly above and to the left of the two-blade curve.

5.4   WIND TURBINE GENERATOR SPEED CONTROL SYSTEMS

The objectives of speed control systems for a wind turbine are to (i) capture maximum energy from the wind; (ii) protect the rotor, generator, and electronic equipment; and (iii) prevent runaway upon loss of electrical load to the generator. There are many methods and designs used to control rotor speed. The method depends on the size of the machine, cost, reliability, and effectiveness of the design. Modern wind machines almost universally use pitch control as the primary method of rotor speed control. Blade pitch offers a wide range of torque control and power transmitted from the wind to the rotor. Maximum power and torque are achieved when the blade pitch is at the ideal lift angle for the instantaneous wind speed and rotor speed. This maximum operating condition is determined by TSR. Blade pitch also may cause stall condition, where power and torque are zero. Precise control of blade pitch can provide the optimum power for wind speeds from cut-in to cutout.

There are five regions of operation for the wind turbine, depending on the wind speed and turbine design. The first is *cut-in*, which is the wind speed where the wind has sufficient power to turn the blades and create measurable power output. Cut-in wind speed may be as low as 3 m/s for small machines to 8 m/s for large machines. The specified cut-in value, as determined by the manufacturer, is when the power in the wind is sufficient to overcome rotor inertia, mechanical losses, and power loss associated with the electronic equipment. If the machine turns on below this threshold, electrical losses in the generator and electronics make it uneconomical to operate (Figure 5.8).

Cutout is the wind speed at which the wind turbine must shut down to prevent damage. The *constant C_p* region is wind speeds between cut-in and rated output. Blade pitch controls operate to maintain maximum TSR and maximum power output at these wind speeds. Since wind speeds are in this range most of the time, the wind turbine is designed to operate at a high coefficient of power to capture maximum energy.

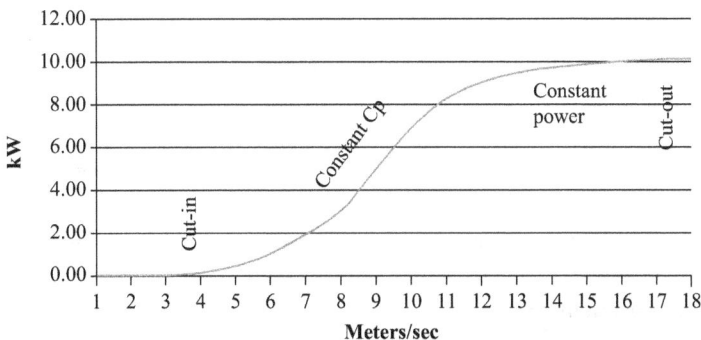

Figure 5.8. WTG power curve.

The *constant power* region is from rated output power at the end of the constant C_p region to cutout. In this range of wind speeds, the wind produces sufficient power to operate the WTG at maximum output. The blade pitch control will stall the blades when wind exceeds this limit. Therefore, the turbine does not operate at the maximum power coefficient or TSR in this region.

The last region is near cutoff, where the wind speed is too high for safe operation. The blade is "feathered" into stall condition in this region, and other types of braking may be applied. Mechanical brakes and electrical eddy current braking to the generator may be used.

5.5   WIND TURBINE INSTALLATION

Ideal land locations for wind turbines are flat areas with strong consistent winds. Prevalent wind direction also benefits the layout of wind turbines in a wind farm. Turbulence caused by upwind turbines will decrease performance of downwind machines. Therefore, the recommended spacing between rows in a wind farm is 8- to 12-rotor diameters. Rows are orthogonal to the prevalent wind direction. Spacing between wind turbines in each row is usually 2- to 4-rotor diameters. This arrangement assumes that the prevalent wind direction is significant, and cross-winds (parallel to the rows) are infrequent. Tower height is a significant factor in performance, since the wind power density is proportional to hub height. Towers are also an expensive component of the turbine installation. Tower sizes of 80 to100 m are common for turbines that are rated from 1 to 3 MW. A rule of thumb is that the tower height must be at least equal to the rotor diameter.

Other installation issues related to wind turbines are noise, electromagnetic interference (EMI), visual impact, and bird and wildlife impacts. Blade and generator noise has been mitigated by smooth airfoil design, nacelle construction, and mechanical improvements. Tubular tower design and slow rotor speed have been shown to reduce bird fatalities. Studies on bird migration have also influenced the location of wind farms. Rotating blades have caused some EMI with local communication systems. This problem may be minimized by proper location of wind turbines and communication towers. The visual impact of wind turbines is of concern to many environmentalists. Many wind farms are located in remote locations, but the large towers and transmission lines associated with the farms are part of the landscape and reflective of the energy use of our society.

5.6 ELECTRIC MOTORS AND GENERATORS

WTGs have incorporated many types of electrical generators over the years. DC generators, alternators, induction motors, synchronous generators, and permanent magnet generators have been used, depending upon the size and application of the wind turbine. Small wind turbines of early design used DC motors that generated voltage and current favorable to battery charging and stand-alone applications. Alternators replaced DC motors because voltage and current output could be controlled to match loads and rotor speed. Early on, induction generators became the preferred interconnected wind turbine due to the rugged simplicity of design, cost, and availability. Synchronous generators are used in larger turbines and require a gearbox to step up the rotor speed to 1,200 or 1,800 rpm. Although synchronous generators run at fixed speeds, electronic AC–DC–AC output systems allow variable speed operation. Doubly fed induction generators are now used in larger turbines to allow variable speed operation and decrease the size of electronic control systems. Permanent magnet generators allow variable speed operation and are used in small- and medium-sized wind turbines. This section reviews the basic construction and operation of these types of machines as WTGs.

5.6.1 DC GENERATOR AND ALTERNATOR

The DC generator produces DC at a specific voltage range determined by the windings in the armature. The stator is the outer shell of the machine with windings on steel cores that produce the electromagnetic field. A DC field current is supplied to the stator, usually from a shunt connection from the armature. The rotor (armature) consists of a single coil of n turns wound lengthwise on a steel core along the shaft. The armature coil is connected to insulated bars around the end of the shaft, which makes up the commutator. Carbon brushes ride on the copper bars with spring tension, and connect external output terminals to the armature winding. As the rotor turns, the commutator rectifies the armature current by switching the polarity of the armature winding. Since the output of the DC generator comes from the inside armature, this design is sometimes called "inside-out." Output is controlled by increasing or decreasing the field current in the stator. The main disadvantage of the DC generator is the need for the commutator brushes, which wear out, require maintenance, generate sparks, and add resistance to the output circuit. An advantage of the DC machine is precise speed control and DC output (Figure 5.9).

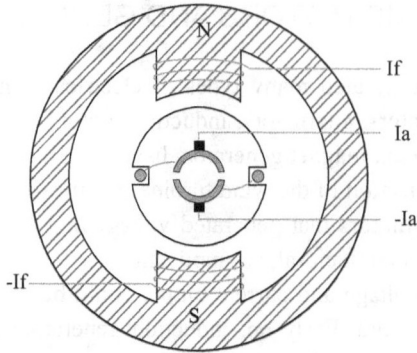

Figure 5.9. Elementary DC machine.

The alternator eliminates the need for the commutator, and uses solid-state rectifiers to convert AC output to DC. The field current is supplied to the rotor (armature) coil through slip ring brushes. As the rotor turns, AC current is generated in the stator windings by the electromagnetic field produced by the rotor. The AC current is rectified to DC with a full-wave solid-state rectifier. Increasing or decreasing rotor field current controls output. This design is used in most automobile battery charging systems. The alternator is used by small wind turbines for DC battery charging, and also in interconnected systems with a line-commutated inverter. The field current requirement can be eliminated by using permanent magnets in the rotor. The permanent magnet machine will be discussed later in this chapter.

5.6.2 THE SYNCHRONOUS GENERATOR

The synchronous machine is used in most utility power plants for system power generation. It runs at a precise speed and has the ability to provide real and reactive power to the grid. It also is well suited to generate three-phase power. The stator is constructed similarly to the alternator and induction machine, with windings symmetrically placed in slots in the steel core. When connected to the grid as a motor, electromagnetic flux is generated at the utility frequency. The rotor, on the other hand, is wound to produce a north–south magnetic field proportional to the field current. The rotating stator flux "pulls" the rotor into synchronism so that the rotor magnetic field locks into the rotating field, and no slip occurs. The rotor axis remains within a few degrees of the stator field axis. Any deviation

Three phase 2-pole generator

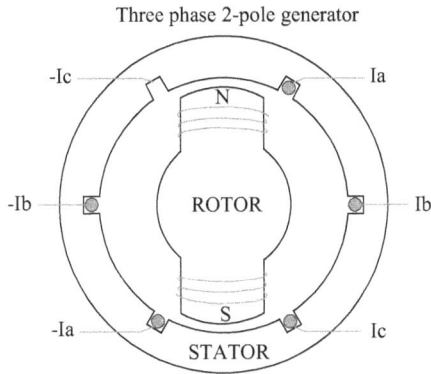

Figure 5.10. The two-pole synchronous generator.

Synchronous 4p 3ph generator

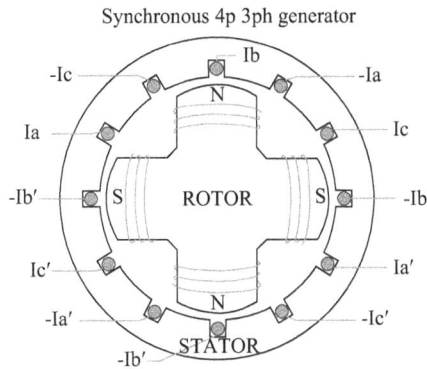

Figure 5.11. The four-pole synchronous generator.

from the synchronized position creates a large opposing torque until equilibrium is again maintained (Figure 5.10).

Synchronous speed is determined by the number of poles (P) on the stator and the frequency of the utility-interconnected line (60 Hz).

$$N_s = 120 \times f/P \qquad (5.21)$$

A two-pole machine (one-pole pair) has a synchronous speed of 3,600 rpm. The synchronous speed is 1,800 rpm for a four-pole machine, and 1,200 rpm for a six-pole machine (Figure 5.11).

As a generator, the synchronous machine must be brought up to the speed by the prime mover. When the rotor speed matches the stator

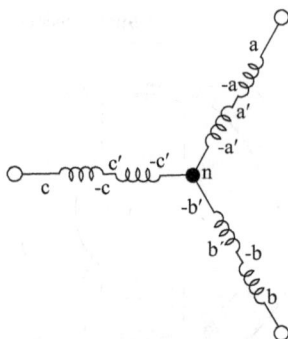

Figure 5.12. Synchronous
generator stator winding.

field frequency, the stator winding contacts are closed. Increasing the
field current maintains synchronism and is also used to control reactive
components of power output. Since the speed is fixed, the synchronous
machine does not match the variable speed aspects of a wind turbine. A
DC–AC–DC electronic converter is usually installed on the stator output
to allow variable speed operation when connected to the fixed utility
frequency grid.

The stator for a four-pole, three-phase synchronous machine is connected in wye as shown in Figure 5.12. Two windings for each phase are
connected in series. The polarity of the windings must be wired such that
voltage is added for each phase. One end of each second winding is connected to a neutral point, which may be connected to the utility system
neutral. Connections may also be made in delta configuration if the neutral
is unnecessary.

5.6.3 THE INDUCTION GENERATOR

The induction generator eliminates two problems with the DC and synchronous machines: brushes and a separate field source. The stator poles
are windings embedded in slots and wound for single- or three-phase
application. Output of the induction motor is the stator winding instead
of the armature in the DC machine. The rotor does not have slip rings or
a commutator, but is constructed of parallel conducting bars around the
rotor shaft with end rings. The end rings short the bars on each end. When
power is applied to the stator, it acts as a transformer, inducing current in
the rotor bars. The rotating flux in the stator brings the rotor up to speed
determined by the frequency of the AC stator current. Torque is generated

Three phase induction generator

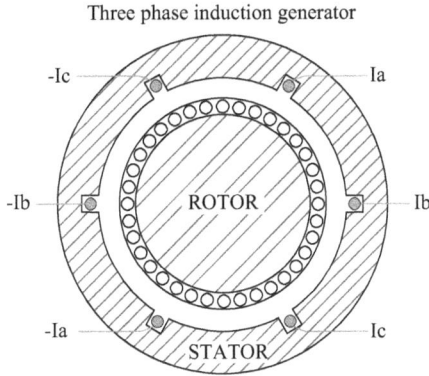

Figure 5.13. The induction generator.

from the flux created in the air gap between the rotor and the stator. If the rotor turns at the same speed as the rotating flux generated by the stator, no air gap flux is generated and no current flows in the rotor. As shaft load is applied and the rotor slows down, current flows in the rotor and torque is generated in the air gap, keeping the rotor speed (N_r) near the rotating flux speed (N_s) in the stator. The simple, rugged design of the "squirrel cage" induction motor makes it ubiquitous in the electric industry.

Synchronous speed of an induction machine is determined by the number of poles wound into the stator. If the machine is three-phase, the three phases represent one-pole pair. The generator in Figure 5.13 shows one set of windings for each phase or one-pole pair. Synchronous speed is calculated as in Equation 5.22, or $Ns = 120 \times 60/2 = 3,600$ rpm. Slip is the difference in rotor speed (N_r) and synchronous speed, expressed as a ratio.

$$S = (N_s - N_r)/N_s \qquad (5.22)$$

If a four-pole induction motor runs at 1,750 rpm, the slip is 0.0278 (2.8%).

Induction motors become generators when torque is applied to the rotor shaft and the rotor speed is above synchronous speed (negative slip). If the four-pole induction motor is connected to a wind turbine rotor through a gearbox, the torque of the blades increase the speed of the rotor until it becomes a generator. If the rotor speed is 1,850 rpm, the slip becomes −0.0278 or −2.8%. If the induction machine is connected to the utility grid, it will feed power into the grid. If the grid connection opens and the wind turbine continues to apply torque, the machine will overspeed and may cause damage. The stator field and flux are necessary

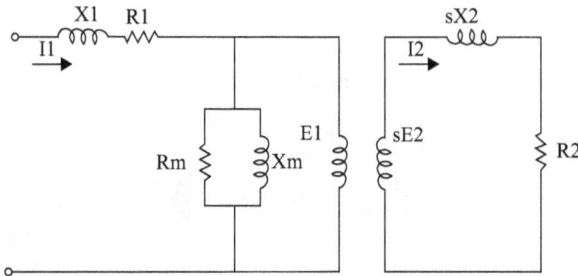

Figure 5.14. Induction machine equivalent circuit.

to operate an induction machine, as the flux will decay rapidly without a source. An emergency brake is needed to prevent unloaded operation. The loss of grid connection and induction generator power is also a safety factor for the utility. Utility standards require that a nonutility generator stop producing power when the grid fails. Tests have shown that an induction machine will continue to produce power if capacitors are connected to the stator terminals. However, the resulting power is unstable in frequency and voltage, and may easily be detected with frequency relays in the wind turbine control system. The induction machine must be supplied with reactive power, and will lower the power factor of the utility system.

The equivalent circuit [24] developed for the induction machine is similar to that of a transformer. R_1 and X_1 are impedance elements of the stator (primary), R_2 and X_2 are rotor impedance components (secondary), and a magnetizing circuit is R_m and X_m. Slip (s) is the equivalent of the transformer–turns ratio (Figure 5.14).

Three-phase conversion power for the induction machine is given by:

$$P = 3\ I_2^2 R_2\ (1 - s)/s\ \text{W} \tag{5.23}$$

It is important to note that power is proportional to the rotor resistance R_2. If the rotor is at zero slip (synchronous speed), I_2 is 0 and power is 0. At standstill, slip is 1 and power is 0. Torque is also proportional to R_2, and is given by $T = P \times \omega$ where ω is angular velocity.

5.6.4   THE WOUND ROTOR INDUCTION GENERATOR

The induction generator may also be constructed as a *wound rotor* machine. Such a machine is not as rugged and simple as the squirrel cage design, but adds a couple of important characteristics to operation. Slip

wound rotor induction generator

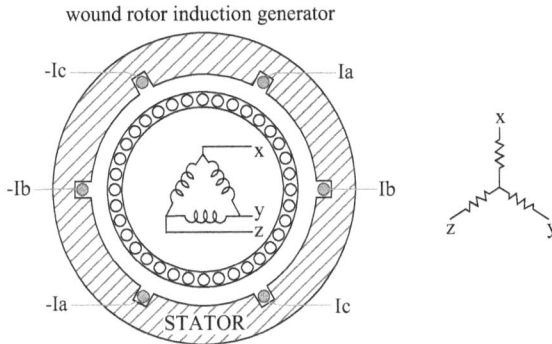

Figure 5.15. The wound rotor induction machine.

rings are installed on the rotor shaft to allow access to the rotor windings. Resistance is added to the rotor circuit windings for improved starting torque characteristics and variable speed operation. The wound rotor is commonly used in streetcars and locomotives because of the high starting torque characteristic. A rheostat in the rotor circuit allows variable speed operation. It has also found use in wind turbines because of the variable speed characteristic (Figure 5.15).

The *doubly fed induction generator* uses a voltage source frequency converter to inject slip frequency power to the rotor. The result is a generator that runs in synchronism with the utility interconnection. Unlike the synchronous machine, only 20% to 30% of the output power goes through the frequency converter. The doubly fed generator has better startup characteristics. One disadvantage of the large doubly fed induction machine is its tendency to drop out upon loss of utility power or a voltage dip. Utilities require that wind turbines ride through temporary outages to prevent a blackout if a large machine drops offline. Capacitors and control circuits have been developed to avoid this problem.

5.6.5 THE PERMANENT MAGNET GENERATOR

As mentioned earlier in this chapter, an alternator may be constructed with permanent magnets in the rotor instead of a field winding. The magnetic field created by the magnets creates the air gap flux and output current in the stator, which is then rectified by solid-state equipment. Permanent magnet generators are used in small and some medium-sized (50 to 100 kW) wind turbines.

The permanent magnet generator is constructed with rare-earth magnets in the rotor. The magnets provide the magnetic field required to induce voltage in the stator. The advantage of this design is the elimination of slip rings that bring current into the rotor for synchronous, DC, and alternator machines. A disadvantage is the expense and weight of these magnets. Larger machines in the range of 100 to 200 kW are now being designed with permanent magnets, which have proven to be more efficient and durable.

PROBLEMS

Problem 5.1

What is the specific wind power density (SP) of a potential wind site at sea level with an average wind speed of 12 m/s? What is the SP if the site is at a 2,000 m elevation?

Problem 5.2

If a horizontal-axis wind turbine with a rotor diameter of 20 m is installed at the 12 m/s wind site and at 2,000 m elevation, what is the net power at this wind speed if the rotor coefficient of power (C_p) is 0.4?

Problem 5.3

What is the net power output of Problem 5.2 if a vertical-axis wind turbine is installed with a diameter of 10 m and height of 20 m?

Problem 5.4

What is the generator rating and blade length for a 1,500/80 wind turbine? What is the SRC of this wind turbine?

Problem 5.5

A10 kW wind turbine is designed to produce rated output at 14 m/s wind speed, a generator-rated speed of 1,200 rpm, with a gearbox ratio of 6:1, and a blade radius of 5.135 m. What is the TSR at rated speed and output?

CHAPTER 6

ENERGY STORAGE AND STAND-ALONE SYSTEMS

Storing energy has always been a challenge for renewable energy systems. Since renewables are generally intermittent, a means of storing energy is critical to stand-alone systems and interconnected renewables. Conventional power systems must provide electrical power when renewable energy is not available. This causes stability problems for the utility and reduces investment recovery. Electrochemical storage (battery) is the prevalent means for electrical storage, but many other methods are becoming practical and economical. Flywheels, compressed air, superconducting coils, and hydrogen fuel cells are a few other methods being developed. Efficiency of conversion from thermal and mechanical energy to electrical energy has limited the development of most of these alternatives. This chapter will review battery storage methods and briefly discuss the alternatives.

Electricity is a highly ordered form of energy, which means that it is easily converted to thermal or mechanical energy. However, thermal energy is a disordered form of energy and the conversion to electricity is inefficient. For example, conventional thermal conversion (power plant) is less than 50% efficient. The process of burning fossil fuels to drive generators and create electricity is inefficient, although it is the primary source of all electricity generated. Unfortunately, photovoltaic (PV) cell and wind turbine technology is less efficient, although growth and technological advances are lowering costs. If an efficient method of energy storage were available, renewable sources could replace centralized power plants with local generation. Transmission lines could be minimized, pollution reduced, and the negative effects of fossil fuel combustion mitigated.

6.1 BATTERIES

Battery storage using electrochemical means is a semiordered form of energy. Electricity generated from battery storage is easily converted to heat or light, but the conversion process within the battery is relatively inefficient. Battery types are categorized as primary and secondary. *Primary batteries* are nonreversible—they cannot be recharged and are discarded after the energy is consumed. Common alkaline cells are examples of this type. *Secondary batteries* are rechargeable. Lead–acid batteries are the most common type and used in automobile and backup systems. Efficiency of a secondary battery is 70% to 80% for the round-trip (charge and discharge) cycle. Energy is lost in the form of heat for both the charge and discharge cycle. Other common types of secondary batteries include nickel–cadmium (NiCad), nickel–metal hydride (NiMh), lithium-ion (LI), zinc–air, and lithium–polymer.

6.1.1 LEAD–ACID BATTERIES

The lead–acid battery consists of a lead (Pb) anode, lead oxide (PbO_2) cathode, and sulfuric acid (H_2SO_4) as the electrolyte. The acid requires a plastic case, and other materials such as antimony or calcium are added to the plates for added strength and performance. Deep cycle batteries used in renewable storage systems are sometimes called "traction batteries" and use thicker plates and greater plate surface area than the typical auto battery. The typical auto battery is sometimes called an "ignition" or "starter" battery, because it is designed for short high-current duty. The classic lead–acid battery is commonly referred to as "flooded" if the electrolyte is in liquid form. Recombinant technology allows lead–acid batteries to be sealed and eliminates the need to add water or vent hydrogen. The batteries are referred to as "valve regulated lead–acid." The lead–acid battery is also available as a sealed "gel-cell" version, which has a nonspillable electrolyte and can be mounted in any position. The *gel-cell* battery cannot be charged or discharged at high rates. The gel may "bubble" at high temperatures, causing permanent voids and reduced capacity. However, the advantages of the gel-cell battery have caused it to become a common replacement for flooded batteries.

State of charge (SOC) is a measure of the remaining capacity of the battery. The most accurate method is measurement of the specific gravity (SG). A fully charged lead–acid battery has an SG of about 1.28, depending on temperature. SG is the density ratio of the sulfuric acid electrolyte compared to water. Charge and discharge measurements with appropriate *C/D* efficiency factors based on battery temperature also provide a good estimate of SOC. SOC may also be estimated by measuring cell voltage— for example, 11.4 V for a 12 V battery (6 cells at 1.9 V/cell) indicates about 20% SOC.

The lifetime of a lead–acid battery is dependent on the number and depth of discharges (DOD). *Depth of discharge* is the percentage of capacity that has been removed compared to the fully charged state. DOD is the complement of SOC. If the battery is discharged by 80% of full capacity, the DOD is 80% and the SOC is 20%. DOD is usually limited to between 75% and 80% to protect from freezing and extend life. The lead plates eventually combine with the acid to create lead sulfate, leaving the capacity of the remaining plates too low to serve the load.

An improvement to the gel-cell lead–acid battery is the *absorbed glass matt* (AGM) design. A porous matt composed of boron-silicate contains the sulfuric acid electrolyte between the plates. The matt is 95% saturated with acid, and sometimes called "starved electrolyte." The matt prevents spillage, eliminates freezing damage, eliminates water loss, and has a low self-discharge rate. However, the cost of AGM batteries is two or three times that of flooded lead–acid batteries.

6.1.2 NICKEL–CADMIUM BATTERIES

NiCad, or nickel–cadmium, uses a cadmium anode and nickel-hydroxide cathode in a stainless steel case. The electrolyte is potassium hydroxide (KOH) with a nylon separator. NiCad cells are almost half the weight of lead–acid cells. The sealed cell means no spilling or mounting issues. Two problems with the NiCad battery have caused it to be less popular. Cadmium is an environmentally dangerous metal, and disposal is restricted. Cells also exhibit a memory problem, causing reduced capacity before reaching expected life. Full discharge to zero capacity and recharge must be performed regularly to prevent memory failures.

6.1.3 LITHIUM-ION BATTERIES

LI cells are the preferred battery for computer systems, because of their high energy density capacity. The atomic weight of lithium is 6.9, compared to 207 for lead. LI cell voltage is 3.5 compared to 2.0 for lead–acid. Fewer cells for the same application voltage mean lower manufacturing costs. However, the plates must be thicker to maintain adequate life, which increases cost significantly for larger capacity batteries. The LI battery also requires an elaborate charging circuitry to prevent damage by overcharging.

6.1.4 NICKEL–METAL HYDRIDE BATTERIES

NiMh cells are similar in construction to NiCad cells except for the metal hydride anode. NiMh batteries are also used for many low-power applications, although they exhibit high self-discharge, and must be charged regularly to maintain capacity. NiMh also has poor peak power capacity and is susceptible to overcharging damage.

6.1.5 OTHER TYPES OF BATTERIES

Other types of batteries being developed are lithium–polymer and zinc–air. The polymer battery shows good specific energy characteristics, but the solid electrolyte limits capacity. The zinc–air battery also displays good specific energy characteristics, but requires adequate air ventilation for the oxygen to carbon and zinc electrode exchange.

Specific energy (watts-hours/mass) is a measure of performance when weight is a primary concern in the application. *Specific density* (watt-hours/volume) is of primary importance when space is critical. Figure 6.1 shows specific energy and specific density for various types of rechargeable batteries. The lead–acid battery has the lowest energy density and lowest specific energy of all current electrochemical types. It is also the least costly. Lithium batteries have the highest densities but also are more expensive.

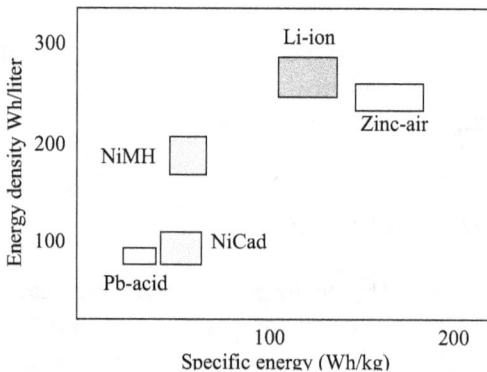

Figure 6.1. Specific energy comparison.

6.2 CHARGE AND DISCHARGE EFFICIENCY OF BATTERIES

The charge efficiency for lead–acid batteries is high when the battery is at a low SOC and low at high SOC. If charging current remains high when the battery approaches full charge, gassing, and heat loss occur. Excess hydrogen gas generation is a safety concern, and excessive heat may damage the battery plates. An effective charger reduces charging current to a trickle when approaching full charge.

Internal resistance increases significantly at low temperatures, which means low discharge amperage and capacity. Figure 6.2 shows the effect of temperature on the capacity of a lead–acid battery. Below zero temperatures significantly decrease capacity, which explains why a cold auto battery delivers low current and poor starting ability.

Any battery typically requires more ampere-hours (Ah) of charge to restore to full SOC after a specified Ah discharge. The charge discharge ratio (C/D) is the Ah input over Ah output for zero net change in SOC. If C/D is 1.1, that means the battery needs 10% more Ah charge than what was discharged to restore to the fully charged state. Battery specifications should provide the C/D ratio for a range of temperatures, as shown in Figure 6.3.

DOD and the number of cycles determine the life of a battery. The life cycle of a NiCad battery is shown in Figure 6.4 for a range of temperatures. Higher temperature decreases the life of a battery. If the battery is discharged to 50% of its capacity, it will fail quicker than if the DOD is limited to 30% of capacity.

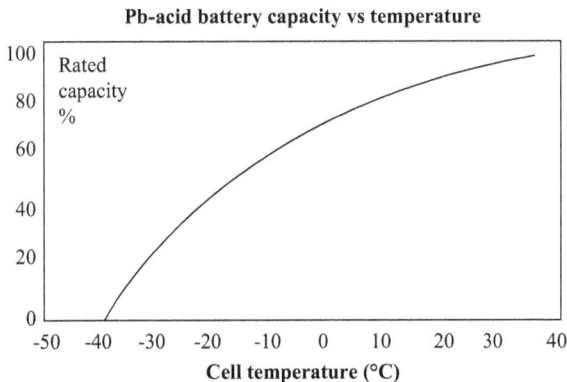

Figure 6.2. Battery capacity and temperature.

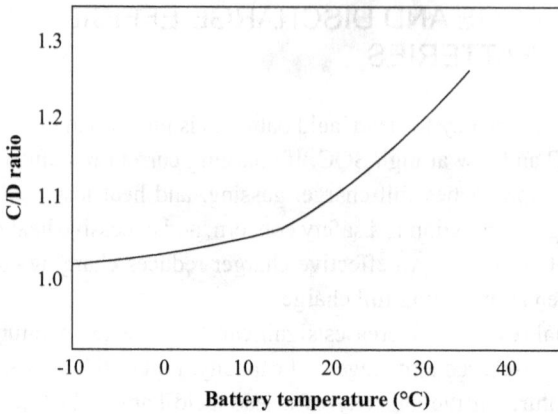

Figure 6.3. Temperature and charge/discharge ratio.

C/D life cycle of NiCad battery vs temp and DoD

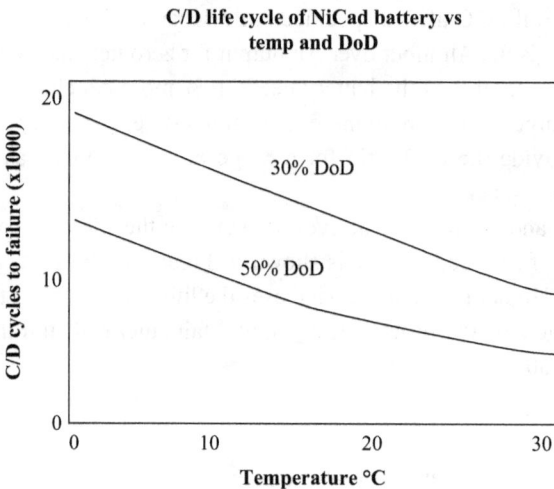

Figure 6.4. Life cycle and DOD.

A common specification for the capacity of batteries is the amp-hour rating (C) at a given charge/discharge rate. *C/n* represents the charge or discharge rate in amperes (A) and charge/discharge time (*n*) in hours.

C/20 is a common rating for deep cycle batteries. For example, a 100-Ah C/20 battery will provide 100 Ah of energy at a 5 A rate for 20 h. If the 100-Ah battery is discharged at a higher rate (e.g., 10 A), it will probably provide less than 100 Ah of energy in 20 h. Likewise, the 100-Ah battery will deliver more than 100 Ah if discharged at a lower current (e.g., 2 A).

A $C/10$-rated 100-Ah battery will deliver 100 Ah if discharged at 10 A, but will probably fall short of the rated capacity when discharged at 20 A. It is important to check the capacity specifications and select a battery that meets the requirements of the load. The charge controller must also match the C/n specification for optimum performance and to protect the battery from over- or undercharging.

Battery failure occurs if a series cell opens, shorts, or experiences capacity loss. If a parallel cell fails and creates an open circuit, it is difficult to notice. Also, a weak series-connected cell is difficult to detect. The best method to detect cell failure is to measure SG. If cells can be isolated from the bank, a charge/discharge test may be used. Battery voltage measurement is helpful to measure cell condition, but not as reliable as SG measurement or the discharge test. Frequent battery bank replacement is a good remedy to prevent loss of capacity.

6.3   OTHER STORAGE MEDIUMS

There are many methods to store potential energy for future electrical generation. A ubiquitous example is water storage. There are countless dams and reservoirs constructed to store elevated water that is converted to electrical power through turbines and generators. This section will introduce a few new promising concepts for practical mass storage of energy, including flywheels, superconducting magnets, and compressed air.

6.3.1   FLYWHEELS

The flywheel has been used in a few applications to store energy, but limited by bearing and windage losses. It has been proven to be an effective method of storing kinetic energy in a rotating inertia with innovative designs that reduce losses. The rotor is constructed of a fiber epoxy composite that can withstand high centrifugal forces. Bearings are magnetic, which have very small frictional losses. High-vacuum installations reduce windage losses. Results of these designs have demonstrated flywheel round-trip efficiency of 90%. An advantage of the flywheel is the high discharge life cycle (i.e., greater than 10,000), and a high DOD without the life-cycle losses associated with other storage systems. It is easy to couple with a motor/generator to convert mechanical energy to electrical energy. The only problem left with the development of effective flywheel applications is scaling to meet storage capacity requirements of electric utility applications.

6.3.2 SUPERCONDUCTING MAGNETS

Energy storage in magnetic fields is a developing technology that has been proven at a small scale (8 kW). Resistance decreases as temperature decreases, and approaches zero (superconducting state) as absolute temperatures are approached. This critical temperature is around 9°K for niobium–titanium, which has been extensively used as the conducting element in the superconducting coil. Recently, more types of materials have been found that exhibit superconducting properties at much higher temperatures—around 100°K. This development makes this method of storage much more feasible, as liquid nitrogen cooling may be used instead of helium, and significantly less refrigeration power is required. The charge/discharge efficiency of the superconducting magnet is around 95%. An advantage over the flywheel is no moving parts, and minimal maintenance for a 30 y life.

The basic elements of the superconducting storage system are a cryostat with the superconducting coil, voltage regulator and shorting switch, inverter, transformer, capacitor, and refrigeration system. The working principle is that energy is stored in the magnetic field of a coil. Energy stored in the coil is related to the current and inductance:

$$E = \tfrac{1}{2}\, I^2 L \text{ (joules)} \qquad (6.1)$$

where L is the inductance of the coil
I is the coil current
The relationship between the voltage and current is given by:

$$V = RI + L\,(di/dt) \qquad (6.2)$$

When the resistance (R) of the coil approaches zero, the circuit does not require a voltage to maintain current in the coil and the magnetic field. The coil terminals may be shorted and the energy in the coil "freezes." The current continues to flow in the coil indefinitely. When power is required by the load, the switch opens momentarily to charge a capacitor. The capacitor is connected to the load through an inverter. A small amount of periodic power is required by the charging circuit to overcome losses in the non-superconducting elements of the circuit. Charge and discharge cycles may be very short, making superconducting magnets an attractive alternative for supplying large amounts of power in a short time.

6.3.3 COMPRESSED AIR

Energy storage using compressed air is effective if a large volume is available. Underground cavern storage is preferred to above-ground tanks for safety and economic reasons. Air may be under constant pressure—as below an aquifer or a variable volume tank—or constant volume—as in a storage tank, cavern, depleted oil or gas fields, or abandoned mines. The advantage of constant pressure is that capacity remains constant as the volume is depleted. Generators may continue to run at rated capacity to maximum DOD. A disadvantage of compressed air systems is the increase in the temperature of the air when compressed. As the gas temperature comes down after compression, some of the energy is lost, with a corresponding loss in stored energy. Many compressed air plants have been built around the world with capacities up to 300 MW. The round-trip efficiency is estimated at 50%. Approximately 100 MWh of energy can be stored with 1 million cubic feet of storage at 600 psi. Studies show that the lead–acid battery and pumped water storage are less-expensive alternatives.

6.4 STAND-ALONE SYSTEMS

Stand-alone renewable energy systems can be simple lighting systems, hybrid residential energy systems, or remote village designs. Solar cars are also examples of stand-alone systems. The common feature of a stand-alone system is that it is not interconnected with a utility system. It may be a direct current (DC) or alternating current (AC) system. Another common feature is some type of storage system, usually electrochemical batteries. The economics of stand-alone systems are based on the unavailability of a utility connection. A remote location with the expense of extending utility service may justify a renewable source and storage system. The added cost of batteries and related maintenance usually prohibits installing a stand-alone system if utility service is readily available. This section will explore the design of residential systems using PV or wind energy renewables and lead–acid battery installations, which may be scaled for small or larger systems. Stand-alone designs are significantly more complex than grid-connected systems.

Storage of electric energy is the key element of a stand-alone system. The renewable source by nature is intermittent, and energy must be stored with deep cycle batteries. The capacity of the battery bank maintains the load for a reasonable duration when the renewable source is not available. Otherwise, a fossil fuel backup generation system is added to extend storage capacity. A small diesel or gas-fueled generator may be used to charge the batteries if the designed duration is exceeded and the load is critical. Backup generation adds significantly to the cost of the system, including fuel costs, maintenance, and inefficient operation if not fully loaded. The alternative to backup generation is shutdown of the electrical system. Several hours of blackout may also be a reasonable alternative.

6.4.1 STAND-ALONE SYSTEMS WITH PHOTOVOLTAICS

PVs are frequently used in stand-alone systems as a source for battery charging. A charge controller regulates output of the PV array to maximize energy transfer to the battery. Each module is designed with the proper number of cells to match battery voltage. Series-connected modules are also designed for voltage appropriate for the charge controller and battery bank. For example, a 36-silicon-cell module is typically used to supply 17 V DC for a 12 V battery. The charge controller regulates PV output voltage for the battery bank and acts as the peak power tracker to optimize PV output. The charge controller provides maximum PV current when the battery is at a low SOC, and low current as the battery reaches full charge, preventing damage to the battery bank. A stand-alone PV array will not operate as efficiently as an interconnected array, because the battery and load may not need the full output of the PV system. The battery may be near full charge when the load is minimal, and excess energy is lost. In contrast, a line-commutated inverter always provides maximum load to the PV module.

A stand-alone inverter converts DC battery voltage to AC power at a set frequency (i.e., 60 Hz). The inverter acts independent of the PV and charge controller, but may be integrated into one package. The inverter is sized to the load being served. It is also used to shed load or shut off if the voltage drops below safe battery levels. As described in Chapter 4, inverters may be square wave, modified square wave, or pure sine wave types, depending on the quality of AC power required by the load.

6.4.2 DESIGN PROCEDURE

The first step in the design of a stand-alone system is load analysis. Determine the power required by each load and the estimated duration or run time for each appliance connected to the electrical system. A table should be constructed that shows power and energy requirements for each load at different times of the year.

Two factors are derived from Table 6.1: peak demand and average daily load. The total peak power (demand) in watts is 2,060, but we can assume that all loads will not be on simultaneously. *Load factor* is the ratio of highest expected load to the total connected load. A load factor of 40% gives us an average peak load of 824 W. Peak load determines the capacity of the inverter. An inverter rated for 1,000 continuous watts and 2,000 maximum watts would be a good choice for the load in Table 6.1. Average daily load determines the capacity of the battery. *Autonomy* is the number of days without PV generation that the system will support the expected load. Three days of storage is considered normal backup for a residential system with a standby generator to charge the batteries if the renewable source is not available. Winter months are usually with the highest usage if air conditioning is not included in the load. The table gives us 6,240 Wh/day load, which equals 18.7 kWh for the three-day backup.

The AC voltage for this example is 120 V single phase. The inverter DC input voltage is determined by wire length, battery capacity, and the renewable source voltage. The most common systems are 12, 24, or 48 V. A 12 V system is usually the best fit for a small PV system, because 36 cell modules are readily available. 12 V batteries are also common. A larger capacity system might use 48 V DC to reduce DC current and wire size. If we choose 24 V DC for the load in Table 6.1, the required amp hours is 18,720 Wh/24 V = 780 Ah. The next step is choosing a battery that provides the voltage and Ah rating required by this load.

Batteries may be purchased as individual 2 V cells, 3-cell 6 V units, or 6-cell 12 V units. Batteries must be wired in series to obtain the system DC voltage, and in parallel to obtain the Ah requirement. Two 12 V batteries in series equal the 24 V system voltage. If we choose 200 Ah deep cycle lead–acid batteries, four parallel batteries will provide 800 Ah capacity. The bank will consist of four parallel strings of two 12 V batteries without correction for temperature or DOD limits. If the battery bank is stored in a heated area, no temperature correction is needed. For outdoor storage or minimal heating, a temperature correction factor must be included, which will increase the bank size, or reduce the three-day backup period in the winter. A DOD limit should also be applied to increase the life of the battery. If we limit DOD to 20% times capacity, the capacity must be increased. The derated capacity is 975 Ah (780 Ah/0.8), and two more batteries must be added, for a total of 1,000 Ah capacity. The three-day backup will draw the 10 battery bank down to a SOC of 200 Ah, which will decrease battery life if allowed to occur often.

The renewable system is sized to recharge the battery bank each day with an estimated output for any time of the year. PV system output will be limited

Table 6.1. Load analysis

Load description	Power (W)	Dec.–Feb.		Mar.–May and Sep.–Nov.		Jun.–Jul.	
		Hours per day	Watt-hours per day	Hours per day	Watt-hours per day	Hours per day	Watt-hours per day
Kitchen lights	200	4	800	3	600	2	400
Living room lights	150	2	300	1	150	1	150
Bedroom lights	80	1	80	1	80	1	80
Refrigerator	150	7	1,050	7	1,050	8	1,200
Microwave oven	600	0.5	300	0.5	300	0.5	300
TV, stereo	180	2	360	2	360	1	180
Furnace fan	400	8	3,200	5	2,000	0	0
Washer	300	0.5	150	0.5	150	0.5	150
Total	2,060		6,240		4,690		2,460

in the winter, and a reasonable estimate of output would be the daily peak sun hours (PSH) for December times the PV system size. The average daily peak sun hours (PSH) is 3.5 for Denver, Colorado, in December. 3.5 PSH per day is equal to 3,500 Wh/m²/day, and if we install a 1-kW system, it will produce 3.5 kWh/day on average in December. (We assume that the PV modules are rated at 1,000 W/m² irradiance.) A 2-kW system will produce 7.0 kWh/day on average in December. The load for one winter day is 6,240 Wh, which is less than the energy a 2 kW system will produce. Again, we must check the charge capacity of the battery to ensure that it will absorb the PV output. If the bank is not heated, the charge capacity may be limited to 50%–60% for temperatures below 0°F. Let us assume that the battery is in a protected space, and charge capacity is derated by 10%. The effective energy from the 2 kW PV system used to charge the battery is 10% less than 7,000 Wh, or about 6,300 Wh. The 2 kW PV will barely provide the load requirement for one winter day (6,240 Wh).

6.4.3   STAND-ALONE SYSTEMS WITH WIND ENERGY

A wind turbine generator (WTG) may be used as the renewable source for a stand-alone system similar to the PV application. The system load is calculated with the same methodology used in the PV example. VThe battery bank is designed, as in the example above, for the load capacity and voltage determined by the wind generator output. An inverter will be used to convert battery DC voltage to 120 V AC load voltage.

Since wind generators are typically installed a long distance from the load, and up a 100 to 200 ft tower, the wire size and voltage drop are critical factors. Long wire causes high voltage drop. For this reason, the battery bank voltage should be 48 V or higher. Instead of a charge controller, the wind generator may be an alternator with field winding control. A regulator controls the field current and increases or decreases the charging current depending on the SOC and voltage of the battery. Similar to an auto regulator, the wind turbine regulator reduces current to the battery as the SOC increases by decreasing field current to the alternator. Voltage is limited to about 2.3 V/cell for lead–acid flooded batteries for a normal charge rate.

Sizing the wind turbine depends on the wind resource, load demand, and average daily load. Given the connected load of 2,060 W, a load factor of 40%, and an average peak load of 824 W, the maximum generator capacity should be about 2,000 W. However, we must determine the energy produced by the wind turbine for the average wind speed at the site to properly size the wind turbine. Several methods may be used to select a wind turbine that will generate the energy that meets the load. If the WTG manufacturer supplies a power curve, we can estimate the average power produced for the site wind speed. Another method is to use Equation 5.3 (see Chapter 5) to determine the rotor diameter and select a WTG with appropriate capacity.

To apply the power Equation 5.3, we must calculate air density for the elevation. Equation 5.6 determines an air density (ρ) of 0.986 kg/m³ for an elevation of 2,000 m.

$$\rho = 1.225 \text{ kg/m}^3 - (1.194 \times 10^{-4}) \times 2,000 \text{ m} = 0.986 \text{ kg/m}^3 \qquad (6.4)$$

The power coefficient may be provided by the WTG manufacturer or estimated at 40% for a two- or three-blade turbine. The best method to determine average wind speed is to install a monitoring system at hub height and collect 2 or 3 y data. If actual measurements are not available, data from a local airport may be used. However, we must extrapolate the average wind speed from the airport anemometer height to the WTG hub height using Equation 5.17. Let us assume that the average wind speed for the proposed site is 6 m/s. The average power required to recharge the battery each day is 260 W (6,240 Wh/day/24 h) from the load in Table 6.1. The generator, field regulator (i.e., charge controller), battery efficiency, and wire loss may be approximated in a derating factor. If the derating factor is estimated to be 20%, the average load is 260/0.8 = 325 W. Substituting this information in Equation 5.3 gives us the following relation:

$$P = 325 \text{ W} = \frac{1}{2} (0.986 \text{ kg/m}^3) A (6 \text{ m/s})^3 (0.4) \qquad (6.4)$$

$$A = 2 \times 325/[(0.986) (6)^3 (0.4)] \qquad (6.5)$$

$$A = 7.63 \text{ m}^2 \qquad (6.6)$$

The diameter is 3.12 m (10.3 ft) from the calculated area ($A = \pi D^2/4$). A wind turbine with a rotor diameter of about 3 m will provide the energy needed to recharge the battery each day. Of course, the wind turbine output is dependent on the intermittent wind. High wind and excess energy will be lost, and low wind conditions will rely on the battery capacity to provide load for three days. Oversizing the WTG by 20% is good practice.

A search for wind turbines for home use will provide several wind turbines of rotor diameter of about 10 ft. We can check the predicted annual energy output for these machines to match the load requirements of 6.24 kWh/day. The SW Wind Power Whisper 200 has a 9-ft diameter rotor and predicts 3,005 kWh/y energy output for a 14 mph (6.26 m/s) average wind speed [25]. This equals 8.2 kWh/day, and exceeds our load estimate by about 30%. A Kestrel e300i turbine with a 10-ft diameter rotor predicts an annual energy output of 3,356 kWh/y, and also exceeds our battery recharging needs. If our average wind speed is 12 mph, these wind turbines predict annual outputs of 2,254 and 2,551 kWh/y, which are close to our load requirement. Both of these machines are designed to charge a battery bank of 12, 24, or 48 V.

6.4.4   HYBRID STAND-ALONE SYSTEMS

Hybrid systems integrate several types of renewable energy sources. The most common system combines a gas or diesel generator with PVs, a wind turbine, and a battery. The advantage of a hybrid system is an increase in the certainty of meeting the demand load at all times. Wind and solar irradiation is intermittent, but the likelihood of having one or the other source for the three-day backup period is much higher than that with one source. The conventional gas or diesel backup system can be minimized or eliminated.

Hybrid systems are more complex to design than simple stand-alone systems with single sources because the voltage and power must match the battery capacity and load. If a conventional generator provides AC to the load, the renewable systems must be synchronized, requiring special monitoring and control equipment. If all sources are available, the battery charge controller must be sized to limit power to the battery charging capacity. Also, excess energy is wasted and reduces the overall efficiency of the system. Power control systems are available that integrate the battery, renewable energy systems, and conventional systems using transfer switches and protection devices.

6.4.5   FUEL CELLS AND HYDROGEN

The fuel cell has potential as a stand-alone and hybrid component. The fuel cell has many advantages over conventional power generation that uses fossil fuels. The most favorable attribute is zero emissions. A hydrogen-powered fuel cell produces water and no airborne pollutants. Fuel cell plants are replacing the diesel generator and battery uninterruptible power supply (UPS) systems. They also are being used to meet peak demands of utility systems. Fuel cells have several disadvantages that limit their application. They are more expensive than conventional power systems. Hydrogen fuel is also expensive and difficult to store and transport. Hydrogen may be generated with renewable systems using electrolysis, a very energy-intensive process.

Several countries and energy companies are working to replace fossil fuels with hydrogen as a primary fuel. Italy constructed the Enel power plant, which is the first industrial-scale facility with a capacity of 20 MW. Enel receives hydrogen as a byproduct from a petrochemical plant and plans to use refuse-derived fuel from solid waste. On a large scale, the city of Reykjavik, Iceland, is planning to become the first city powered totally with hydrogen by 2050. The country is converting transportation means to hydrogen, and building hydrogen fuel centers.

The fuel cell is an electromechanical device that generates electricity by a chemical reaction that does not change the electrodes or electrolyte materials. The electrodes do not wear out, as in batteries, and there are no moving parts, which give the fuel cell a long, maintenance-free life. The energy-producing process is the reverse of electrolysis. Hydrogen and water combine in an isothermal operation to produce electricity and water. If natural gas, ethanol, or methanol are used to produce hydrogen, carbon dioxide and some traces of carbon monoxide, hydrocarbons, and nitrous oxides are by-products, although less than 1% of that produced by the internal combustion engine [26].

There are several types of fuel cells with various energy densities and costs. Power density ranges from 0.1 to 0.6 W/cm^3. The phosphoric acid fuel cell is the most common in relatively small applications. The alkaline fuel cell (AFC) is used by aerospace systems and is the most expensive type. The AFC uses KOH as the electrolyte with porous anodes and cathodes. The proton-exchange membrane is also known as the solid-polymer fuel cell and is the most promising type for small-scale applications, such as home power systems. The molten carbonate fuel cell works at higher temperatures (600°–700°C), and may be the choice for utility-scale applications in the future. The higher-temperature cells are the most efficient. Solid-oxide fuel cells are another high-temperature design that may be used in large utility or wind farm applications. Costs range from $200 to $1,500 per kW, which is 2–15 times the cost of diesel engines.

Hydrogen storage is a problem for fuel cells, vehicles, and all hydrogen fuel systems. Compressed H_2 gas is notorious for developing leaks, and liquid hydrogen requires high compression and expensive storage tanks. Liquid H_2 is best for long-term and large-quantity storage. On the bright side, hydrogen has a much higher energy density than storage in an electrochemical battery. Gaseous hydrogen can store about 47,000 BTU/ft^3, compared to 2,000 BTU/ft^3, for a conventional battery. Liquid hydrogen will store up to 240,000 BTU/ft^3. Gasoline is still the most convenient energy storage medium with an energy density of 1,047,000 BTU/ft^3.

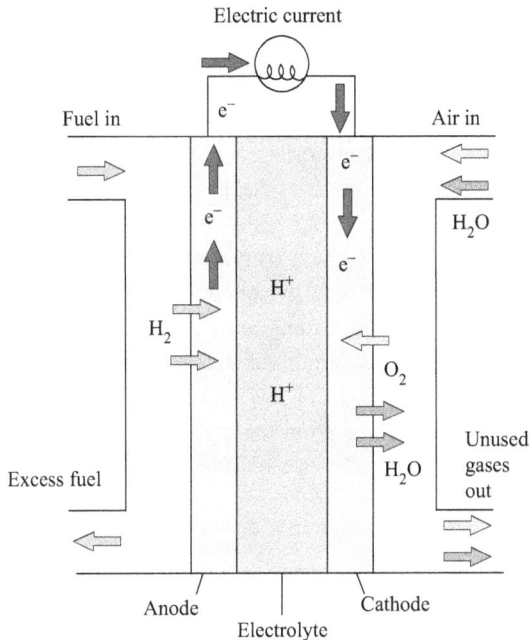

Figure 6.5. Basic fuel cell.

Hydrogen is the simplest and most abundant element. It is composed of a single proton and electron. Electrolysis is a simple method to obtain hydrogen from water. Hydrogen and oxygen gases are produced by passing electrical current through a salt water solution. A fuel cell is the reverse of this process, as electrical current is generated when a fuel and water combine in a cell with an anode, cathode, and electrolyte. Figure 6.5 shows a basic fuel cell.

6.4.6 THE SOLAR CAR

The solar car is the ideal stand-alone design. With PV cells integrated into the car body as the battery-charging system, high-energy-density batteries, and efficient electric motors, the solar car is capable of traveling hundreds of miles under ideal solar conditions. Energy storage with electrochemical batteries is a limitation on the size and reliability of the solar car. The national and international solar races have produced "proof-of-concept" vehicles that are reliable and efficient. However, the efficiency of solar cells and electrochemical batteries are a cost barrier to conventional vehicle application.

Plug-in solar cars do not have PV cells as a renewable source of energy unless the cars are recharged by a solar power plant. The range of plug-in vehicles is limited to about 100 miles and restricts their application to commuting use. Improvements in battery performance are creating a resurgence of this type of transportation. Charging stations are increasing as battery performance improves, which is driving the industry to build more electric vehicles.

The hybrid car has become a common design for commuting vehicles. Without solar cells, the energy source comes from internal combustion engines that recharge the battery and assist driving power under high load or long distance. Many larger vehicles are classified as hybrid vehicles, although the batteries are a low percentage of the energy source. Most of the power is derived from the conventional combustion engine. However, the hybrid car reduces emissions, especially for short drives, and frequent stops.

PROBLEMS

Problem 6.1

A battery array is connected to a 24 V, 10 A load. If the array consists of eight 6 V, 200 Ah capacity $C/20$ lead–acid batteries, how long will the battery array supply the load before a DOD of 50% is reached?

Problem 6.2

Design a battery array that will maintain 4 days of autonomy in a cabin with the following loads and appliances: refrigerator, gas furnace (fan), 200 W of lighting, and miscellaneous loads of 300 W. The load factor is 30%. Use 12 V batteries and a 48 V DC to 120 V AC inverter. Battery SOC should not drop below 40%.

Problem 6.3

Design a PV charging system that will accommodate the battery and loads in Problem 6.2. Assume that the location is Boulder, Colorado. The PV array is a tracking system like the Zomeworks mechanical tracker.

Problem 6.4

Design a small wind turbine charging system that will accommodate the battery and loads in Problem 6.2. Assume that the location is in Wyoming where the average yearly wind speed is 16 mph, and elevation is 2,000 m. What will be the necessary swept area and blade length of the wind turbine?

CHAPTER 7

Economics: Break-Even and Return on Investment

Any capital investment should be analyzed for economic feasibility. "Going green" is not the only reason for investing in a renewable energy system. There are many tangible and intangible benefits associated with renewables. We must look at the big picture and evaluate the economics as well as the externalities. Chapter 1 addressed some of the externalities associated with conventional electric energy generation but did not attempt to place a cost for pollution, global warming, and other societal effects of fossil fuel combustion. As we study the basic economics of renewable energy investment and returns, a dollar amount will not be assigned to externality costs because it is controversial and arbitrary. Our simple economic analysis will provide only a basic reference of what a system costs and what the future returns are in dollars.

7.1 SIMPLE PAYBACK

Payback or break-even (BE) analysis is the most common method used to compare capital investment. Initial investment (I) is recouped by reoccurring revenue (R) over the lifetime of the equipment. Inflation, depreciation, and the time value of money are ignored. The initial investment is the cost of equipment and installation, less incentives, rebates, and tax credits. Although the timing of the tax credits and incentives is not the same as the initial payment, a lumped sum at time zero simplifies the payback calculation. Revenue is also lumped into annual payments instead of the actual monthly utility savings and revenue. Revenue for small systems is usually an estimate of annual energy production at a fixed utility rate.

An example of a simple payback calculation is described in Table 7.1 for the purchase of a 2.4 kW PV interconnected residential system. The

Table 7.1. PV system costs and benefits

System information	Value
System size (watts) DC at (STC)	2,410
Estimated annual production (kWh)	3,249
Estimated first-year revenue ($.11/kWh)	$357.39
System cost	
Equipment and installation	$12,050
Rebates and incentives	
Rebate ($2.35/W)	$5,664
ITC (30% federal tax credit)	$1,915
Net cost after incentives	**$4,471**

DC, direct current; STC, standard test conditions; ITC, investment tax credit.

initial cost of equipment and installation is $12,050. The net cost ($I$) after utility rebate and federal investment tax credit (ITC) is $4,471. Revenue after the first year (R) is estimated as $357.39, using the PVWatts™ program and a local utility rate of 11 cents/kWh (Table 7.1).

Assuming fixed revenue equal to the first-year energy savings, BE in years is calculated as the ratio of investment cost to annual return:

$$BE = I/R = \$4,471/\$357.39 = 12.5 \text{ y} \qquad (7.1)$$

The simple payback analysis ignores many significant factors, such as inflation, rate increases, maintenance, and possible equipment failures. However, it provides a very useful comparison to similar installations at various locations. Average irradiation affects annual revenue. Different utility jurisdictions have different avoided-costs rates for renewable energy. Tax exemptions vary from state to state. Rebates vary widely among utilities, and change with policy.

7.2 ADVANCED BE ANALYSIS

With the help of a spreadsheet, the BE analysis can be improved by adding rate increases, inflation, or PV degradation factors. Table 7.2 shows the BE calculation with a 5% increase in utility rates each year after installation.

Table 7.2. Payback with annual 5% electric rate increases

Milestones	Year	Rate ($)	Savings ($)	Accumulated ($)
After first year of operation	1	0.110	357.39	357.39
	2	0.116	375.26	732.65
	3	0.121	394.02	1,126.67
	4	0.127	413.72	1,540.40
	5	0.134	434.41	1,974.81
	6	0.140	456.13	2,430.94
	7	0.147	478.94	2,909.87
	8	0.155	502.88	3,412.76
	9	0.163	528.03	3,940.78
System cost recovered	10	0.171	554.43	4,495.21
	11	0.179	582.15	5,077.36
	12	0.188	611.26	5,688.62
	13	0.198	641.82	6,330.44
	14	0.207	673.91	7,004.36
	15	0.218	707.61	7,711.96
	16	0.229	742.99	8,454.95
	17	0.240	780.14	9,235.09
	18	0.252	819.14	10,054.23
	19	0.265	860.10	10,914.33
Utility contract complete	20	0.278	903.11	11,817.44
	21	0.292	948.26	12,765.70
	22	0.306	995.68	13,761.38
	23	0.322	1,045.46	14,806.84
	24	0.338	1,097.73	15,904.57
Expected life of system	25	0.355	1,152.62	17,057.19

BE with rate increases is 10 y for the PV system described above. The calculation is continued through the expected life of the system (25 y), and provides an estimate of the total accumulated revenue.

The spreadsheet analysis allows other factors to be included in the evaluation. For example, PV cells degrade with time. The performance warranty is typically 90% of initial output after 25 y of operation. Table 7.3 adds a degradation factor of 0.5%/y. A production-based incentive (PBI) of $0.09/kWh for 10 y is also included (see Chapter 1 for a description of the PBI rebate). The utility energy rate is based on $0.12/kWh in the first year and increases 5%/y. The initial system cost is $14,861 for a 5.5 kW system, and payback is 9 y.

Table 7.3. Payback with inflation and PV degradation

Year	Solar energy (kWh)	Utility rate per kwh ($)	Energy savings ($)	PBI payments ($)	Total savings ($)	Accu-mulated savings ($)
1	7,139	0.120	856.68	642.51	1,499.19	1,499.19
2	7,103	0.126	895.02	639.30	1,534.31	3,033.50
3	7,068	0.132	935.07	636.10	1,571.17	4,604.67
4	7,032	0.139	976.91	632.92	1,609.83	6,214.51
5	6,997	0.146	1,020.63	629.76	1,650.39	7,864.89
6	6,962	0.153	1,066.30	626.61	1,692.91	9,557.80
7	6,927	0.161	1,114.02	623.47	1,737.49	11,295.30
8	6,893	0.169	1,163.87	620.36	1,784.23	13,079.52
9	6,858	0.177	1,215.96	617.25	1,833.21	14,912.73
10	6,824	0.186	1,270.37	614.17	1,884.54	16,797.27
11	6,790	0.195	1,327.22		1,327.22	18,124.49
12	6,756	0.205	1,386.61		1,386.61	19,511.10
13	6,722	0.216	1,448.66		1,448.66	20,959.77
14	6,689	0.226	1,513.49		1,513.49	22,473.26
15	6,655	0.238	1,581.22		1,581.22	24,054.47
16	6,622	0.249	1,651.98		1,651.98	25,706.45
17	6,589	0.262	1,725.90		1,725.90	27,432.36
18	6,556	0.275	1,803.14		1,803.14	29,235.49
19	6,523	0.289	1,883.83		1,883.83	31,119.32
20	6,490	0.303	1,968.13		1,968.13	33,087.45

PBI (production-based incentive)

Other factors that will improve BE analysis are maintenance costs and inflation. Although inflation is difficult to pin down due to recessions and upturns, a moderate rate of 2% or 3% would improve the reality of our calculations. Maintenance costs may include equipment replacement—the inverter warranty is typically 10 y. A lump sum in years 10 and 20 for a new inverter may be appropriate. If the investor is not the homeowner and a business can write off depreciation, an annual depreciation and tax recovery amount should be included. These additional factors are left for the reader to explore.

7.3   RETURN ON INVESTMENT

Another method of economic analysis is return on investment (ROI). ROI determines the rate at which the investment is recovered, and therefore, must have a definite investment period. We can choose a period of 10, 20, or 25y, depending on the investor's preference. If two or more investment opportunities are being evaluated, the investment periods must be equal. Let us choose a 20-y-period and assume that the asset is fully depleted, so the salvage value is zero. Again, let us start with simple ROI that does not include inflation, derating factors, or rate increases. For the PV system above, the initial investment cost is $4,471 and the annual revenue is $357.39. The simple ROI is

$$ROI = \$357.30/\$4,471 = 8.0\% \qquad (7.2)$$

Now we can compare the 8% ROI for the solar investment with another energy investment of the same expected life or period (20 y). A spreadsheet may be created to include inflation, derating, factors, equipment replacement, and so on, to the analysis, similar to the BE analysis. However, more complex rate formulas must be used when the annual revenue is not constant. ROI is also not well suited to short investment periods.

7.4   OTHER ECONOMIC EVALUATION METHODS

Economic evaluations become very complex quickly when the timing of revenue does not occur at the end of a period, as we assume in the above examples. It is always a good policy to draw a timeline showing all costs and revenues before calculating BE or ROI. Economic formulas such as present worth, future value, periodic payments, and so on, can be found in engineering economics textbooks. A course on economic analysis and investment is highly recommended for all engineering students.

PROBLEMS

Problem 7.1

Calculate the BE for a 6 kW PV interconnected system. The installation cost is $4/W. The utility retail rate is 12 cents/kWh. The utility also pays $2.00/W upon installation as a renewable energy incentive. The homeowner also is eligible for the 30% ITC after the first year of operation.

- What is the net system cost after incentives? (The ITC applies after the utility incentive has been applied.)
- What is the annual revenue if the homeowner uses all energy produced? (Use the PVWatts program at its default settings.)
- What is the payback or BE in years?

Problem 7.2

What is the ROI for the system of Problem 7.1 if the investment period is 20 y?

Problem 7.3

What is the BE of Problem 7.1 if the utility rate increases 3% each year?

Problem 7.4

Using a spreadsheet analysis, determine the BE of Problem 7.1 with 3% utility rate increases and a PV degradation factor of 1.5%/y.

Problem 7.5

Add an inflation factor of 3%/y to the result of Problem 7.4 and determine BE.

CHAPTER 8

Solar Thermal Systems

Solar thermal systems have been around for many years to provide hot water for domestic, commercial, and industrial use. Solar irradiation heats a collector and transfer fluid, which transfers thermal energy to storage or direct use. The collector may be flat-plate, evacuated tube, parabolic trough, or concentrator with reflecting mirrors (heliostats). Flat-plate collectors are generally used for domestic hot water (DHW) since the temperatures are limited to about 180°F. The evacuated tube is more efficient and can reach between 300°F and 400°F at the header. The concentrator systems (parabolic trough, heliostat, etc.) heat the collector fluid to very high temperatures that can drive generators for electric power or preheat water for the boiler of a conventional fuel plant. Some type of storage system captures thermal energy for use when solar energy is not available. Domestic storage is usually a well-insulated 200- to 1,500-gallon water tank. Industrial systems may use salt brine that stores energy by changing state (i.e., solid to liquid). All thermal systems require a heat transfer and control system, which may be a simple pump and temperature differential controller. This chapter will describe basic domestic solar hot water systems and a few high-temperature applications used to supplement power systems or generate electricity.

8.1 DOMESTIC HOT WATER

Solar water heating is arguably the best application of solar energy. Domestic solar thermal systems can be very basic, requiring a collector, storage, and a method of heat transfer. Small systems installed on residential rooftops can provide all hot water needs for a home if sized properly. Simple systems for DHW are efficient and inexpensive. In areas where freezing is not a problem, a simple solar collector may consist of a shallow collector box with black absorber plates. These *direct systems*

131

(passive) do not require pumps or controllers and rely upon water pressure to move domestic water from the source through the collector to the faucet. *Indirect systems* (active) use pumps to move the transfer fluid through the collector to a storage tank. Domestic water flows through copper coils in the storage tank using domestic water pressure. Domestic water in an indirect system is always isolated from the solar collector loop. A differential temperature controller turns the collector pump on when the collector temperature exceeds storage tank temperature by a set difference (e.g., 5°F to 10°F). The differential controller also switches the pump off when the collector temperature drops to the storage temperature. There are two basic designs for indirect hot water systems: the open-loop or drainback system and the closed-loop pressurized system. Each design has specific requirements for piping and control systems. The open-loop system is simpler and does not require periodic maintenance as closed-loop systems do.

8.1.1   OPEN-LOOP SYSTEMS

The simplest design for the solar collector loop is the open loop, sometimes called the drainback system. The collector loop is open to the atmosphere and relies on gravity to drain the loop. The collector pump moves water from the storage tank to the collector on the roof (i.e., elevated above the storage tank). When the collector fills with water, it falls back to the storage tank through a vertical or sloped pipe. When the pump shuts off, all collector water flows back to the storage tank. A temperature sensor in the collector prevents operation when the collector temperature is below freezing point. This design eliminates the need for antifreeze in the collector loop. It is intrinsically safe from contamination of domestic water, because the collector and heat transfer pipes use domestic water instead of a glycol mix or oil. A copper coil is installed in the tank to heat domestic cold water for the hot water system. Domestic water pressure forces water through the storage tank coil directly to the faucet, or to a (backup) conventional hot water heater that uses gas or electricity (Figure 8.1).

8.1.2   CLOSED-LOOP SYSTEMS

The collector loop for a closed-loop system is pressurized and contains a 50% glycol water mix or an oil transfer liquid (Figure 8.2). Propylene glycol is usually used because it is not as toxic as other antifreeze coolants. An oil transfer liquid is used if very high temperatures are expected,

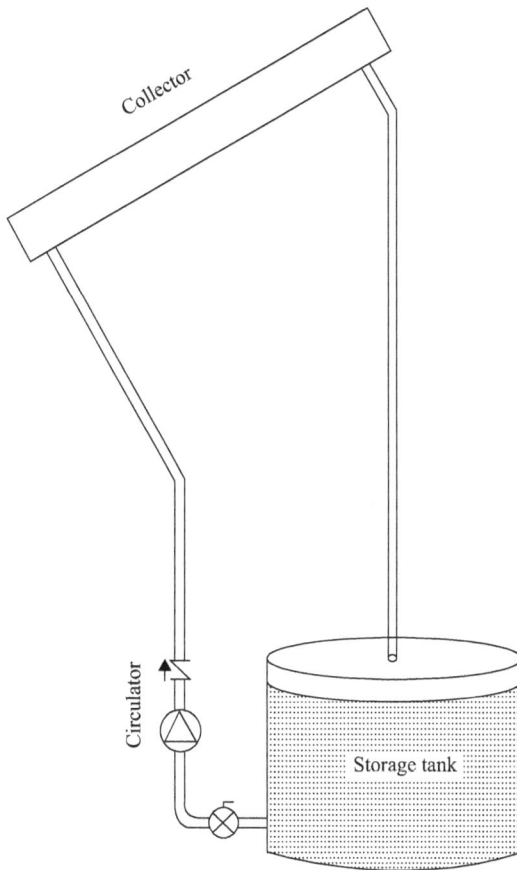

Figure 8.1. Open-loop (drainback) system.

because the boiling point is much higher than a glycol mix. Specific heat capacity for oil is about 0.5, lower than that of water or glycol. The closed-loop system is necessary if the collector cannot be located directly above the storage tank. It also eliminates the possibility of frozen pipes if a freeze switch fails or if the pipes move and no longer drain with gravity. Specific heat capacity for the glycol mix is 0.85, slightly less efficient than the open-loop design, but this difference is negligible for small systems. A copper pipe is used instead of a cross-linked polyethylene pipe (XLPE) in the collector loop due to high temperatures. XLPE is used in domestic water systems and some radiant loops if the fluid temperature is less than 150°F. XLPE is limited to 80 psi and 150°F. The heat exchanger in the storage tank is generally 100 ft of 0.75 in soft copper coiled inside a large (>1,000 gallons) storage tank.

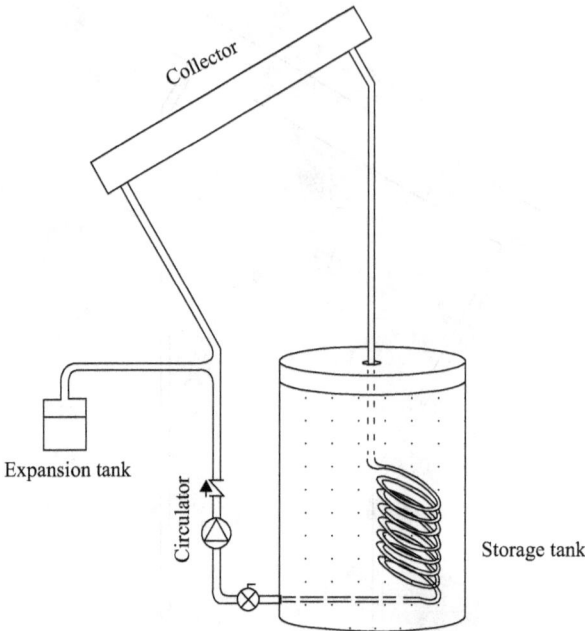

Figure 8.2. Closed-loop system.

8.1.3 COLLECTORS

Flat-plate collectors are commonly used for DHW systems (Figure 8.3). The flat-plate collector is a very durable, proven product that has some advantages over new designs. It can be stagnated (no fluid flow) in full sun without damage. It also sheds snow better than a vacuum tube, because the glass cover will warm from convection within the collector frame. The glass envelope of a vacuum tube remains at ambient temperatures due to the excellent insulation, and will not melt snow or ice. Therefore, vacuum tubes should be installed at a more vertical angle than a flat-plate collector (>50° tilt). The vacuum tube collector (Figure 8.4) generates much higher fluid temperatures, and is more efficient because the vacuum insulates the glass from the atmosphere and minimizes heat loss from the collector plate.

The *vacuum tube* collector takes up slightly more space than a flat-plate because of the separation between tubes. If a single tube in a 30-tube collector fails, it may be replaced individually. Stagnation is also damaging to the vacuum tube collector plates, as the extreme heat build-up will cause plate emissions that collect on the inside glass, reflecting radiation from the

Figure 8.3. Flat-plate collector.

Figure 8.4. Vacuum tube collector.

tube. Vacuum tube systems usually include dual-collector pumps with an uninterruptable power supply backup to ensure coolant flow whenever the differential controller is "on." The flat-plate collector has header pipes at the top and bottom, and the vacuum tube array has one header pipe at the top where each tube delivers heat from a single transmitter tube.

8.2 SOLAR THERMAL SPACE HEATING SYSTEMS

Many domestic solar thermal systems have expanded to include capacity for household space heating in addition to hot water systems. A typical home might have 200 sq. ft of collector and a storage tank of 1,500 gallons. The collector loop is similar to the DHW system using vacuum tube technology and a closed-loop pressurized system. A heat reject pump and radiator are added to dissipate excess energy in summer months when the tank approaches maximum safe temperature (<190°F).

A relatively sophisticated hydraulic system is used to distribute thermal energy from storage to various space heating systems, such as radiant floor heating, furnace fan coils, and energy recovery ventilator equipment. A series primary–secondary loop is shown in Figure 8.5. This design is also called an injector loop system. The primary loop is central to the system with circulator pump P3, which runs whenever a secondary loop needs energy from storage or the collectors. Closely spaced T-fittings connect secondary loops to the primary loop. Figure 8.5 also shows a bypass loop (P7 and HX2), which brings collector energy directly to the primary loop when solar energy is available and there is a call for heat energy by one of the secondary systems. Heat exchangers HX1 and HX2 isolate the glycol loops from the primary loop and storage tank. Since the storage tank is open to atmospheric pressure, heat exchanger HX1 isolates the pressurized collector loop from the storage tank, and HX3 isolates the primary loop from the storage tank. A gas boiler provides stage 2 backup heating. Whenever solar energy cannot keep up with the demands of the house, the gas boiler is started and pump P12 delivers 180°F water to the main loop. The gas boiler also provides backup for DHW with pump P10. Pump P13 provides hot water for the DWH tank when the solar tank is above 120°F.

The primary–secondary loop design prevents flow in one circuit from interfering with flow in another circuit. This is called *hydraulic separation*, and can be designed with series or parallel loops, or with a hydraulic separator tank. Figure 8.5 is an example of the series primary–secondary type. Critical in this design is the distance between secondary tees located in the primary loop. The distance between the injector tees cannot be greater than four times the pipe diameter for proper fluid flow and separation.

Figure 8.6 is an example of a parallel primary–secondary system with a boiler as the primary energy source. The boiler tank may also be heated with solar energy. The secondary circuits are connected into the primary loop with parallel pipes between headers. This design is more complicated than the series system, but may adapt better to the physical layout of the system. It is equally as efficient for separation as the series design. The parallel loop system is usually filled with water, but a glycol mix may be used if any part of the system is exposed to freezing conditions. Heat exchangers may also be inserted in secondary circuits that are exposed to freezing conditions, but an additional circulator pump is then required to circulate glycol in the isolated loop.

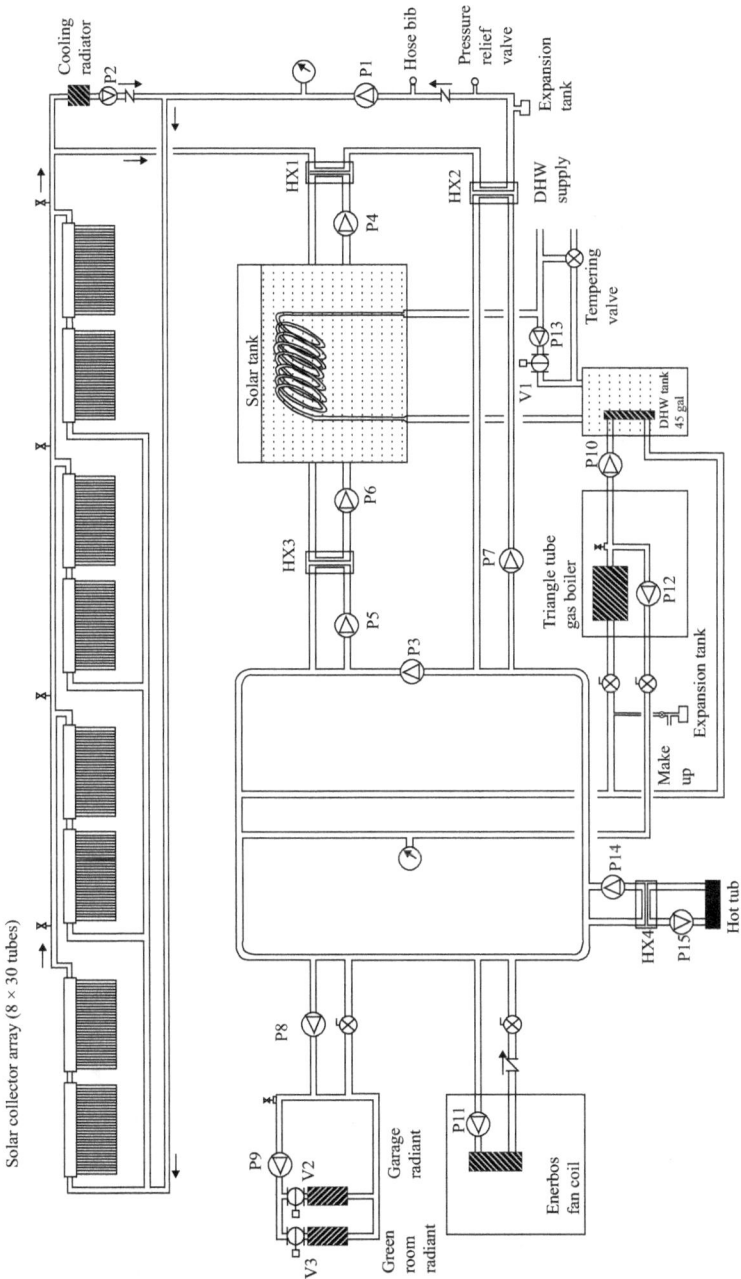

Figure 8.5. Injector loop system.

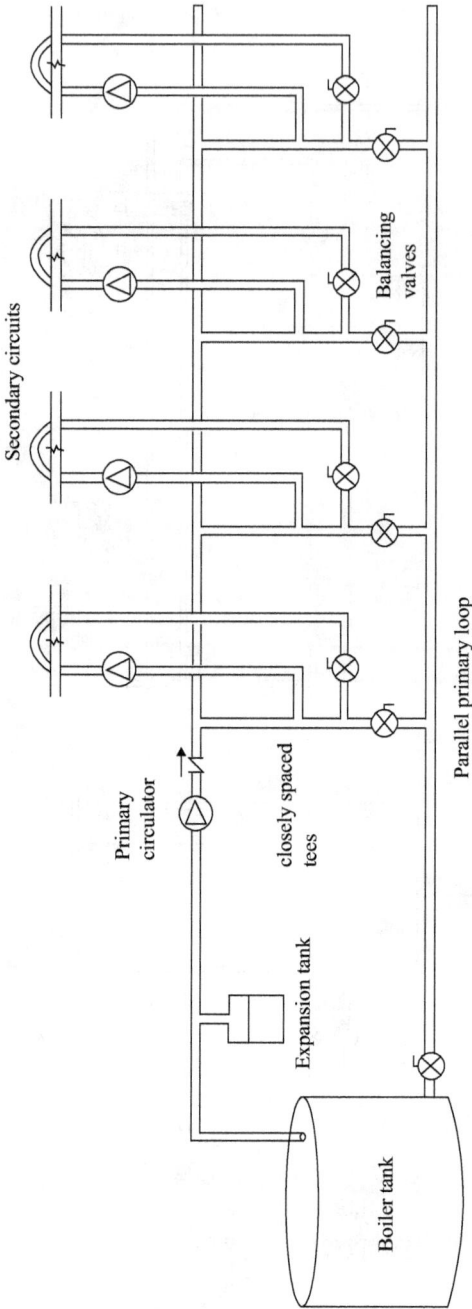

Figure 8.6. Parallel primary–secondary loop design.

8.3 ELECTRIC GENERATION USING SOLAR THERMAL SYSTEMS

Utility-scale solar thermal systems require direct sunlight, and the U.S. Southwest locations are ideal for this technology (Figure 8.7). Indirect or diffuse irradiation will not generate enough heat for efficient operation. Parabolic trough collectors are used when very high temperatures are needed to drive a turbine and electric generator. A 64 MW solar plant (Nevada One) uses parabolic collectors to generate steam for the turbine generator. The parabolic collector has replaced the heliostat mirror system formerly used by Solar One and Solar Two installed near Barstow, California. Solar One operated from 1981 to 1995 and consisted of 1,818 heliostats focused on a 10 MW thermal electric generator. Solar Two operated from 1995 to 2009 and used 1,926 heliostats. It also incorporated a molten salt storage system for round-the-clock generation. The salt used for change-of-state storage is 60% sodium nitrate and 40% potassium nitrate (Figure 8.7).

Xcel Energy installed a solar thermal system at the coal-fired Cameo plant near Palisade, Colorado, to preheat feed water to the boilers used for electric generation in 2010. The solar system is integrated with Unit #2, a 49 MW steam turbine. The thermal system covers 6.4 acres with 8 rows of 500-foot parabolic concentrator collectors. The total collector area is 70,400 sq. ft. The trough is made of curved, pure glass mirrors that focus sunlight on the receiver tube. The transfer fluid is a food-grade mineral oil that is pumped through the tube to a heat exchanger. The operating tem-

Figure 8.7. Solar thermal plant.

perature of the collector is between 350 and 575°F. The heat exchanger heats condenser water from 360°F to 407°F for the boiler feed-water input. A 3% to 5% increase in heat rate efficiency is expected. The solar system will reduce coal use by 900 tons and CO_2 production by 2,000 tons.

8.4   GROUND SOURCE HEAT PUMPS

Many utilities are offering incentives to install ground source heat pump (GSHP) systems in domestic and commercial applications. A homeowner can benefit by eliminating the need of a gas or oil furnace, and rely totally on an electric backup furnace with the GSHP. The GSHP can maintain indoor temperatures using the ground thermal reservoir as a source for most of the year. The GSHP electric load is a pump that circulates water or glycol in the ground collector, and a fan to distribute the heat in a forced-air-handling system. The electric furnace is needed for the coldest days of the year, and an evaporative cooler may be needed for a few hot summer days. The utility also benefits with a slight increase in electric use and reduction of gas demand. A typical domestic application requires a pipe grid installed 20 ft below the surface or vertical pipe drilled to 100 or 200 ft.

PROBLEMS

Problem 8.1

How many British Thermal Units (BTUs) are required to heat a 40-gallon water thermal storage tank from 40°F to 120°F?

Problem 8.2

If the collector flow rate is 2 gallons per minute (GPM), how long must the pump run to generate the BTUs calculated in Problem 8.1? (Assume that the collector increases the water temperature by 20°.)

Problem 8.3

If a vacuum tube collector increases water temperature of a 1,000 gallon storage tank by 40°, about how long did the pump run if the flow rate was 4 GPM?

CHAPTER 9

Structural Considerations for PV Arrays

Photovoltaic (PV) arrays are built to withstand the elements normally experienced at the specific installation location. It is economically unfeasible to build support systems that can withstand all possible catastrophic events at any location. For example, we would not need a system to be designed to withstand hurricane-force winds if the location is far inland and has never experienced a hurricane. The probability of a tornado or an earthquake must be evaluated for each site. Designers use local building codes that are created from historical data. This chapter will focus on design for the most probable events in the Colorado area—high winds and snow loading.

Wind speed maximums are based on historical data recorded at nearby weather stations. Typical data are specified as average values and 3 s gust maximums. The 3 s gust data are the basis for roof and PV system design. Snow load is a critical load factor when located in mountainous or northern areas. The maximum expected accumulation of snow on a horizontal surface is used to calculate dead weight loading on roof-mounted arrays and ground-mount systems. The PV array, roof structure, and structural equipment must be designed to support the additional load due to snow accumulation. This chapter presents a step-by-step procedure to meet the predicted loading conditions.

Step 1: Research local building codes
Step 2: Determine the surface area of the array
Step 3: Calculate the dead loads (equipment and snow loads)
Step 4: Calculate the live loading (wind lift and downforce)
Step 5: Calculate the compressive and tension load per footing
Step 6: Evaluate the strength of fastening components

9.1   RESEARCH BUILDING CODES

Criteria for building construction may come from city, county, or state jurisdictions. Usually, this information can be found online through the jurisdiction's website. Roof-mounted and ground-mount systems require structures
built to meet maximum expected snow loading and 3 s wind gusts. The
criteria may vary by location within the jurisdiction, and maps are available
that specify expected wind speed and snow loading. Typical (3 s gust) wind
speeds in the front range of Colorado are between 110 to 120 mph and snow
load maximums are 20 to 30 lb./sq. ft. In the following example, we will
use snow loading of 25 lb./sq. ft and 110 mph wind speed in our design.

9.2   DESIGN PROCEDURE

This section will demonstrate an engineering procedure used to determine
the design and equipment necessary to meet the structural requirements.
All dynamic and static forces on the array are calculated based on wind
and snow loading. Strength analysis of materials is obtained from the
manufacturer. Spans between footings are determined using manufacturer
specifications for wind and snow loading. Components of the racking system are analyzed for structural tension and compression strength.

9.2.1   DETERMINE SURFACE AREA

The total surface area of a PV array must be known to determine loading on
the structure. For example, if we are installing 24 PV modules with dimensions of 39 × 64 in, the total surface area is 415 sq. ft (17.3 sq. ft × 24 modules).
    The effective wind area is the smallest area of continuous modules.
This is the smallest tributary (contributing load) to a support or to a single-span rail. A single-span rail is a length of rail with supports near the
ends of the rail section. The span is the distance between supports. Figure
9.1 shows a portion of the Unirac Loading table 2. The effective wind
area column provides a range of 10–100 sq. ft. If the smallest area of
continuous modules exceeds 100 sq. ft, use 100 sq. ft.

9.2.1.1   Determining Effective Wind Area

The effective wind area column in the Unirac Wind Loading Table
provides four or five options for each zone. This is the smallest area of
continuous modules that make up the array. In other words, it is that area

Table 2. p_{net30} (psf) Roof and Wall

Basic Wind Speed (mph)

Zone	Effective Wind Area (sf)	90 Downforce	90 Uplift	100 Downforce	100 Uplift	110 Downforce	110 Uplift	120 Downforce	120 Uplift	130 Downforce	130 Uplift	140 Downforce	140 Uplift	150 Downforce	150 Uplift	170 Downforce	170 Uplift
Roof 0 to 7 degrees																	
1	10	5.9	-14.6	7.3	-18.0	8.9	-21.8	10.5	-25.9	12.4	-30.4	14.3	-35.3	16.5	-40.5	21.1	-52.0
1	20	5.6	-14.2	6.9	-17.5	8.3	-21.2	9.9	-25.2	11.6	-29.6	13.4	-34.4	15.4	-39.4	19.8	-50.7
1	50	5.1	-13.7	6.3	-16.9	7.6	-20.5	9.0	-24.4	10.6	-28.6	12.3	-33.2	14.1	-38.1	18.1	-48.9
1	100	4.7	-13.3	5.8	-16.5	7.0	-19.9	8.3	-23.7	9.8	-27.8	11.4	-32.3	13.0	-37.0	16.7	-47.6
2	10	5.9	-24.4	7.3	-30.2	8.9	-36.5	10.5	-43.5	12.4	-51.0	14.3	-59.2	16.5	-67.9	21.1	-87.3
2	20	5.6	-21.8	6.9	-27.0	8.3	-32.6	9.9	-38.8	11.6	-45.6	13.4	-52.9	15.4	-60.7	19.8	-78.0
2	50	5.1	-18.4	6.3	-22.7	7.6	-27.5	9.0	-32.7	10.6	-38.4	12.3	-44.5	14.1	-51.1	18.1	-65.7
2	100	4.7	-15.8	5.8	-19.5	7.0	-23.6	8.3	-28.1	9.8	-33.0	11.4	-38.2	13.0	-43.9	16.7	-56.4
3	10	5.9	-36.8	7.3	-45.4	8.9	-55.0	10.5	-65.4	12.4	-76.8	14.3	-89.0	16.5	-102.2	21.1	-131.3
3	20	5.6	-30.5	6.9	-37.6	8.3	-45.5	9.9	-54.2	11.6	-63.6	13.4	-73.8	15.4	-84.7	19.8	-108.7
3	50	5.1	-22.1	6.3	-27.3	7.6	-33.1	9.0	-39.3	10.6	-46.2	12.3	-53.5	14.1	-61.5	18.1	-78.9
3	100	4.7	-15.8	5.8	-19.5	7.0	-23.6	8.3	-28.1	9.8	-33.0	11.4	-38.2	13.0	-43.9	16.7	-56.4
Roof >7 to 27 degrees																	
1	10	8.4	-13.3	10.4	-16.5	12.5	-19.9	14.9	-23.7	17.5	-27.8	20.3	-32.3	23.3	-37.0	30.0	-47.6
1	20	7.7	-13.0	9.4	-16.0	11.4	-19.4	13.6	-23.0	16.0	-27.0	18.5	-31.4	21.3	-36.0	27.3	-46.3
1	50	6.7	-12.5	8.2	-15.4	10.0	-18.6	11.9	-22.2	13.9	-26.0	16.1	-30.2	18.5	-34.6	23.8	-44.5
1	100	5.9	-12.1	7.3	-14.9	8.9	-18.1	10.5	-21.5	12.4	-25.2	14.3	-29.3	16.5	-33.6	21.1	-43.2
2	10	8.4	-23.2	10.4	-28.7	12.5	-34.7	14.9	-41.3	17.5	-48.4	20.3	-56.2	23.3	-64.5	30.0	-82.8
2	20	7.7	-21.4	9.4	-26.4	11.4	-31.9	13.6	-38.0	16.0	-44.6	18.5	-51.7	21.3	-59.3	27.3	-76.2
2	50	6.7	-18.9	8.2	-23.3	10.0	-28.2	11.9	-33.6	13.9	-39.4	16.1	-45.7	18.5	-52.5	23.8	-67.4
2	100	5.9	-17.0	7.3	-21.0	8.9	-25.5	10.5	-30.3	12.4	-35.6	14.3	-41.2	16.5	-47.3	21.1	-60.8
3	10	8.4	-34.3	10.4	-42.4	12.5	-51.3	14.9	-61.0	17.5	-71.6	20.3	-83.1	23.3	-95.4	30.0	-122.5
3	20	7.7	-32.1	9.4	-39.6	11.4	-47.9	13.6	-57.1	16.0	-67.0	18.5	-77.7	21.3	-89.2	27.3	-114.5
3	50	6.7	-29.1	8.2	-36.0	10.0	-43.5	11.9	-51.8	13.9	-60.8	16.1	-70.5	18.5	-81.0	23.8	-104.0
3	100	5.9	-26.9	7.3	-33.2	8.9	-40.2	10.5	-47.9	12.4	-56.2	14.3	-65.1	16.5	-74.8	21.1	-96.0
Roof >27 to 45 degrees																	
1	10	13.3	-14.6	16.5	-18.0	19.9	-21.8	23.7	-25.9	27.8	-30.4	32.3	-35.3	37.0	-40.5	47.6	-52.0
1	20	13.0	-13.8	16.0	-17.1	19.4	-20.7	23.0	-24.6	27.0	-28.9	31.4	-33.5	36.0	-38.4	46.3	-49.3
1	50	12.5	-12.8	15.4	-15.9	18.6	-19.2	22.1	-22.8	26.0	-26.8	30.2	-31.1	34.6	-35.7	44.5	-45.8
1	100	12.1	-12.1	14.9	-14.9	18.1	-18.1	21.5	-21.5	25.2	-25.2	29.3	-29.3	33.6	-33.6	43.2	-43.2
2	10	13.3	-17.0	16.5	-21.0	19.9	-25.5	23.7	-30.3	27.8	-35.6	32.3	-41.2	37.0	-47.3	47.6	-60.8
2	20	13.0	-16.3	16.0	-20.1	19.4	-24.3	23.0	-29.0	27.0	-34.0	31.4	-39.4	36.0	-45.3	46.3	-58.1
2	50	12.5	-15.3	15.4	-18.9	18.6	-22.9	22.2	-27.2	26.0	-32.0	30.2	-37.1	34.6	-42.5	44.5	-54.6
2	100	12.1	-14.6	14.9	-18.0	18.1	-21.8	21.5	-25.9	25.2	-30.4	29.3	-35.3	33.6	-40.5	43.2	-52.0
3	10	13.3	-17.0	16.5	-21.0	19.9	-25.5	23.7	-30.3	27.8	-35.6	32.3	-41.2	37.0	-47.3	47.6	-60.8
3	20	13.0	-16.3	16.0	-20.1	19.4	-24.3	23.0	-29.0	27.0	-34.0	31.4	-39.4	36.0	-45.3	46.3	-58.1
3	50	12.5	-15.3	15.4	-18.9	18.6	-22.9	22.2	-27.2	26.0	-32.0	30.2	-37.1	34.6	-42.5	44.5	-54.6
3	100	12.1	-14.6	14.9	-18.0	18.1	-21.8	21.5	-25.9	25.2	-30.4	29.3	-35.3	33.6	-40.5	43.2	-52.0

Figure 9.1. Wind load table. (*Source*: Courtesy of Unirac, Code Compliant Installation Manual, 2010, table 2.)

of the fewest number of modules on a run of rails. For example, if two modules are installed apart from the main array, the effective wind area becomes the area of those two modules.

9.2.1.2 Roof Zone

The Unirac Wind Loading Table provides five possible zones in which the array is installed. Zone 1 is the area within a 3-foot perimeter of the roof edges. This is the area that is least susceptible to the effects of the wind. Zones 2 and 3 encroach on the areas close the roof edges, and Zone 4 is in the most susceptible areas. Zone 5 is a vertical (wall) installation. Drawings and descriptions of roof zones can be found in the *Unirac Code-compliant Installation Manual* online at unirac.com.

9.2.2 CALCULATE DEAD LOADS

The maximum expected dead weight on the PV array due to snow loading (S) is 10,380 lb. (25 lb./sq. ft $\times$ 415 sq. ft). Additional dead weight on the structure includes equipment weight. If each PV module weighs 41 lb., total weight due to modules is 984 lb. (24 modules $\times$ 41 lb.). Other balance-of-system (BOS) equipment includes rails and mounting brackets. The estimated weight due to BOS equipment may be estimated per linear foot of rails. A typical value is 2 lb./linear ft, and the 24-module array requires 90 ft of rails. BOS dead load is then 180 lb. Total dead weight due to equipment (D) is 1,164 lb. (984 lb. + 180 lb.).

9.2.3 CALCULATE DEAD AND LIVE LOADS DUE TO WIND FORCES

Wind loading consists of two components: downforce (Wd) and uplift (Wu). These components depend on several factors, including roof height, array tilt, roof zone of installation, topography, and exposure area. These factors can be obtained from organizations that build structures for PV installations. Unirac provides a very useful code-compliant manual [27]. Tables in the manual include values for Wd and Wu given the aforementioned factors (see Figure 9.1). For example, given a roof height of 20 ft, Zone 1 installation (PV array within a 3-ft perimeter of roof edges), topographic factor of 1.0, roof height adjustment factor of 1.29, and tilt of 20°, the effective downforce is 8.9 lb./sq. ft, and the effective uplift is –18.1 lb./sq. ft. Factoring in the adjustment factor of 1.29 with the effective area of 415 sq. ft, total downforce (Wd) is 4,765 lb. (8.9 lb./sq. ft $\times$ 415 sq. ft $\times$ 1.29), and total uplift is –9,690 lb. (–18.1 lb./sq. ft $\times$ 415 sq. ft $\times$ 1.29).

Now we can calculate the total downforce (Pd) on the array, which includes all dead weight loads and wind downforce.

$$Pd = D + S + Wd = 1,164 + 10,380 + 4,765 = 16,309 \text{ lb.} \tag{9.1}$$

Total uplift (Pu) is the uplift due to maximum wind speed.

$$Pu = Wu = -9,690 \text{ lbs} \tag{9.2}$$

9.2.4 CALCULATE COMPRESSION AND TENSION LOADING

Each footing for the support structure must support part of the total loading on the array. If we have 40 footing brackets and assume all footings

Roof Pitch 3:12 to 4:12 (14.04° to 18.43°)

v	0 psf	10 psf	20 psf	30 psf	40 psf	60 psf	80 psf	100 psf
				Snow Load				
85 mph	10.0 ft	8.5 ft	6.5 ft	6.0 ft	5.0 ft	3.5 ft	2.5 ft	2.0 ft
90 mph	9.5 ft	8.5 ft	6.5 ft	6.0 ft	5.0 ft	3.5 ft	2.5 ft	2.0 ft
100 mph	8.5 ft	8.5 ft	6.5 ft	6.0 ft	5.0 ft	3.5 ft	2.5 ft	2.0 ft
105 mph	8.5 ft	8.5 ft	6.5 ft	6.0 ft	5.0 ft	3.5 ft	2.5 ft	2.0 ft
110 mph	8.0 ft	8.0 ft	6.5 ft	6.0 ft	5.0 ft	3.5 ft	2.5 ft	2.0 ft
120 mph	7.5 ft	7.5 ft	6.5 ft	6.0 ft	5.0 ft	3.5 ft	2.5 ft	2.0 ft
135 mph	6.5 ft	6.5 ft	6.5 ft	6.0 ft	5.0 ft	3.5 ft	2.5 ft	2.0 ft
150 mph	6.0 ft	6.0 ft	6.0 ft	6.0 ft	5.0 ft	3.5 ft	2.5 ft	2.0 ft

Roof Pitch 5:12 (22.62°)

v	0 psf	10 psf	20 psf	30 psf	40 psf	60 psf	80 psf	100 psf
				Snow Load				
85 mph	11.5 ft	8.5 ft	6.5 ft	6.5 ft	5.5 ft	4.0 ft	3.0 ft	2.5 ft
90 mph	10.5 ft	8.5 ft	6.5 ft	6.5 ft	5.5 ft	4.0 ft	3.0 ft	2.5 ft
100 mph	9.5 ft	8.5 ft	6.5 ft	6.5 ft	5.5 ft	4.0 ft	3.0 ft	2.5 ft
105 mph	9.5 ft	8.5 ft	6.5 ft	6.5 ft	5.5 ft	4.0 ft	3.0 ft	2.5 ft
110 mph	9.0 ft	8.5 ft	6.5 ft	6.5 ft	5.5 ft	4.0 ft	3.0 ft	2.5 ft
120 mph	8.5 ft	8.5 ft	6.5 ft	6.5 ft	5.5 ft	4.0 ft	3.0 ft	2.5 ft
135 mph	7.5 ft	7.5 ft	6.5 ft	6.5 ft	5.5 ft	4.0 ft	3.0 ft	2.5 ft
150 mph	7.0 ft	7.0 ft	6.5 ft	6.5 ft	5.5 ft	4.0 ft	3.0 ft	2.5 ft

Figure 9.2. Rail spans.(*Source*: Courtesy of Renusol, Solar Mounting System Design Guide, 2013.)

are equally loaded, the total tension load per foot is 242 lb. (–9,690 lb./40). Likewise, the compression load per footing is 408 lb. (16,309 lb./40).

Rails must also support the loading on the PV modules. Rail manufacturers usually provide tables that specify the maximum span length for various wind speeds and snow loads. Figure 9.2 is an example of a rail loading table. A range of maximum wind speed is provided in the left column, and the top row provides a range of snow loads. For our 110 mph wind speed and 25 psf snow load, the table allows a span of 6.0 ft if the roof pitch is 3:12 or 4:12. If the roof pitch is 5:12, the span can be 6.5 ft.

Every manufacturer of racking equipment has tables or a method of calculating rail spans. The calculation can be done manually if the bending moment and yield strength for the structural components are known, but the tables are adequate if the manufacturer provides engineering certification.

9.2.5 EVALUATE COMPONENT STRENGTH

The final step of the structural analysis is to determine the strength of the support brackets and calculate a safety factor for each component. We have already used tables provided by the manufacturer to determine rail spans. Mounting brackets usually do not have these convenient tables, but if they are designed for roof-mounted installations, they usually have a sufficient safety factor built into their construction. However, it is always a good idea to check specifications and load test data.

The following analysis is for S-5 brackets, commonly used on steel roof installations. Load test data are given in Figures 9.3 and 9.4. The bracket is fastened to 0.5 in oriented strand board (OSB) for the test given in Figure 9.3, and 0.056 in construction steel for the test in Figure 9.4. Tension tests are done using hydraulic test equipment. Mean yield strength for the OSB test was 844 lb., and 1,659 lb. for the steel plate test. Critical to these test results are the type of bolts used to connect the deck material to the S-5 bracket. Care must be taken to install bolts and lag screws that have equivalent strengths as the mounting brackets. Also, stainless hardware should be used for PV installations, which are subject to corrosive environments. Stainless hardware also prevents electrical corrosion of dissimilar metals for grounding connections.

If the S-5 bracket is installed on a steel roof, the footing bracket's mean yield strength is 1,659 lb., the safety factor for compression is 4.0 (1,659 lb./408 lb.). If the S-5 is attached to OSB, the yield strength is about 844 lb., and the safety factor is about 2. It is critical to determine what type of roof structure is supporting the roof deck. For example, lag screws into wood purlins oriented flatwise will have much less pull-out strength than lag screws into rafters.

The pull-out strength of lag screws used to secure the support brackets to the roof structure (purlins or rafters) must be greater than the uplift force on each footing.

9.2.6 INCORPORATE SAFETY FACTOR

The pull-out strength of lag screws in pine rafters is estimated at 150 lb./in of thread into the wood. For a penetration of 3 in, total pull-out is 450 lb. The safety factor for the lag screw into pine is 1.86 (450 lb./242 lb.). In compression, the safety factor is 1.10 (450 lb./408 lb.). To increase the safety factor, increase the number of footings, and footings should be spaced closer together (i.e., shorter spans).

To complete the structural analysis, all components should be evaluated for structural strength given the loading calculations. L-feet, which are used to connect rails to footings, should have tension and

Customer Name S-5!
Operator James Thomas
Specimen Information 1/2" OSB

Project Number 1003-00416
Provider Information S5
S-5 Product Information CorruBracket CB-
Additional Information Attached w/ four

Load Cell S/N (QS6807), Units (Lbs) 6000
Preload Value (Lbs) 1

Crosshead Speed (Inches / min) or Rate 0.25
Displacement Sensor XHD_100 (XHD100)

Test No	Spec ID	Load @ Yield (lbs)	Maximum Load (lbs)
32914	1	984	984
32915	2	667	667
32916	3	880	880
Mean		844	844
Median		880	880
Std Dev		162	162
Maximum		984	984
Minimum		667	667
Range		318	317

Figure 9.3. Yield strength of S-5 bracket on OSB. (*Source*: Courtesy of Metal Roof Innovations, Ltd. S-5!.)

	Customer Name	S-5!
	Operator	James Thomas
	Specimen Information	.056" Construction Steel

	Project Number	1003-00416
	Provider Information	S5
	S-5 Product Information	CorruBracket CB-
	Additional Information	14x1.5" T3 w/

| | Load Cell S/N (QS68807), Units (Lbs) | 6000 |
| | Preload Value (Lbs) | 1 |

| | Crosshead Speed (Inches / min) or Rate | 0.25 |
| | Displacement Sensor | XHD_100 (XHD100) |

Test No	Spec ID	Load @ Yield (lbs)	Maximum Load (lbs)
32917	1	1,714	1,745
32918	2	1,637	1,757
32919	3	1,626	1,713
	Mean	1,659	1,738
	Median	1,637	1,745
	Std Dev	48	23
	Maximum	1,714	1,757
	Minimum	1,626	1,713
	Range	89	45

Figure 9.4. Yield strength of S-5 bracket on construction steel. (*Source:* Courtesy of Metal Roof Innovations, Ltd. S-5!.)

compression strength greater than or equal to other components. PV module clamps must be designed for the application. A structural engineer should be consulted to determine what safety factors are reasonable and prudent for a secure installation. Finally, all bolts in the installation, especially the module clamps and footing hardware, should be tightened with a torque wrench using specifications provided by the manufacturer (Figures 9.5–9.7).

Figure 9.5. Tilt racking.

Figure 9.6. Roof-mount brackets.

Figure 9.7. Metal roof flush-mounted PV array.

PROBLEMS

Problem 9.1

Determine uplift and downforce values for the 415 sq. ft array if the area wind speed is 120 mph (3 s gusts), and snow loading is 30 psf. The array is mounted on a roof with a pitch of 23°.

Assume the following design factors: roof height of 20 ft, Zone 1 installation (PV array within a 3 ft perimeter of roof edges), topographic factor of 1.0, roof height adjustment factor of 1.29.

Problem 9.2

Given the uplift and downforce values of Problem 9.1, determine footing spans if your footers are fastened into pine wood rafters with 3 in penetration.

APPENDIX A

SHADE ANALYSIS PROGRAMS

Sunnyside Solar

Solar Site Survey Results for: Textbook
  Installation Location: Example 1
  Position: Lat. 39.989899 N, Long. 105.137299 W, Altitude 1627 m
  Weather Data: Reference Station BOULDER, CO.
  Fixed Array: Tilt 22 degrees, Azimuth 180 degrees
  PV Modules: 21 Sanyo Electric, Model: HIP-200BA3
  Inverter: 1 Fronius, Model: IG 4000
  System Derate Factor: 84.6% (cabling, connectors, array soiling, etc)
  Installation Rating (STC): 4.2 kW
  kWh/kWp = 1426
  Electricity Cost: 0.100 $/kWh

Energy Production Estimates:

	Solar Energy (kWh)	DC Energy (kWh)	AC Energy (kWh)	Earned
January	2693.1	399.2	350.0	$35
February	3001.5	435.6	381.0	$38
March	4400.7	634.7	559.2	$56
April	4630.3	651.7	572.3	$57
May	4995.6	690.4	606.4	$61
June	5044.9	682.7	599.9	$60
July	5138.2	681.2	598.0	$60
August	5020.9	667.0	586.3	$59
September	4546.2	613.6	539.6	$54
October	3930.3	546.9	480.9	$48
November	2891.6	421.3	369.0	$37
December	2662.3	396.7	348.0	$35
Total Annual	48955.8	6821.0	5990.6	$599

Figure A.1. Site survey Example 1. (*Source*: Prepared by Sunnyside Solar using Solmetric iPV, version 2.4, page 1 of 2.)

Sunnyside Solar

Solar Site Survey Results for: Textbook

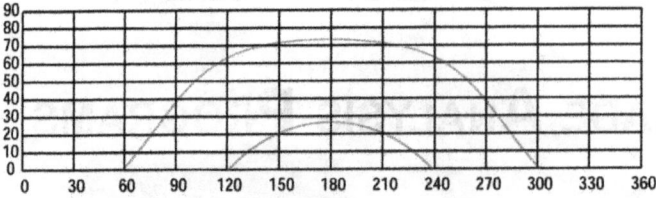

	Solar Energy Actual Tilt: 22 Actual Az:180 No Shade kWh/m sq/day	Solar Energy Actual Tilt: 22 Actual Az:180 Shade kWh/m sq/day	Shade Derating % Captured
January	3.5	3.5	100%
February	4.3	4.3	100%
March	5.7	5.7	100%
April	6.2	6.2	100%
May	6.5	6.5	100%
June	6.8	6.8	100%
July	6.7	6.7	100%
August	6.5	6.5	100%
September	6.1	6.1	100%
October	5.1	5.1	100%
November	3.9	3.9	100%
December	3.5	3.5	100%
Total Annual	5.4	5.4	100%

Figure A.2. Site survey Example 1. (*Source*: Prepared by Sunnyside Solar using Solmetric iPV, version 2.4, page 2 of 2.)

Table A.1. PVWattsprogram output file

PVWatts Program Results — Example 1

Station identification

City:	Boulder
State:	Colorado
Lat (deg N):	40.02
Long (deg W):	105.25
Elev (m):	1634

PV system specifications

DC Rating:	4.2 kW
DC to AC Derating Factor:	0.77
AC Rating:	3.2 kW
Array Type:	Fixed Tilt,
Array Tilt:	22
Array Azimuth:	180

Energy specifications

Cost of Electricity:	8.4 cents/kWh

Results

Month	Solar Radiation (kWh/m²/day)	AC Energy (kWh)	Energy Value ($)
1	3.68	370	$31.08
2	4.30	385	$32.34
3	5.69	558	$46.87
4	6.18	568	$47.71
5	6.44	600	$50.40
6	6.70	588	$49.39
7	6.58	580	$48.72
8	6.45	567	$47.63
9	6.04	526	$44.18
10	5.09	475	$39.90
11	3.90	371	$31.16
12	3.50	353	$29.65
Year	5.39	**5,942**	$499.13

Source: Courtesy of NREL

Table A.2. PV Wattsprogram derate table

Calculator for overall DC to AC derating factor		
Component derating factors	**Component derating values**	**Range of acceptable values**
PV module name plate DC rating	0.95	0.80–1.06
Inverter and transformer	0.92	0.88–0.100
Mismatch	0.98	0.97–0.996
Diodes and connections	1.00	0.99–0.998
DC wiring	0.98	0.97–0.100
AC wiring	0.99	0.98–0.994
Soiling	0.95	0.30–0.996
System availability	0.98	0.00–0.996
Shading	1.00	0.00–1.01
Sun-tracking	1.00	0.95–1.01
Age	1.00	0.70–1.01
Overall DC to AC derating factor	0.77	(PV Watts Default)

Source: Courtesy of NREL online PV Watts Program.

If the default overall DC to AC derating factor of 0.77 is not appropriate for your PV system, you may use this calculator to determine a new value by changing one or more of the component derating factors in the table and clicking the "Calculate Derate Factor" button. You may enter values within the ranges shown in the table. Values outside the ranges are reset to the default values by the derating calculator. The new overall DC to AC derating factor may be hand-entered or copied and pasted into the "DC to AC Derate Factor" field on the PV Systems Specifications section of the PVWatts input form. Click on help below the table for information about DC to AC derating factors.

PHOTOVOLTAIC MODULE SPECIFICATIONS

SANYO

All HIP-xxxBA3 Models

Electrical Specifications		180W	186W	190W	195W	200W	205W
Rated Power (Pmax)[1]	W	180	186	190	195	200	205
Maximum Power Voltage (Vpm)	V	54.0	54.4	54.8	55.3	55.8	56.7
Maximum Power Current (Ipm)	A	3.33	3.42	3.47	3.53	3.59	3.62
Open Circuit Voltage (Voc)	V	66.4	67.0	67.5	68.1	68.7	68.8
Short Circuit Current (Isc)	A	3.65	3.71	3.75	3.79	3.83	3.84
Minimum Power (Pmin)	W	171.0	176.7	180.5	185.3	190.0	194.8
Max System Voltage (Vsys)	V	600	600	600	600	600	600
Series Fuse Rating	A	15	15	15	15	15	15
Temperature Coefficient (Pmax)	%/°C	-0.33	-0.30	-0.30	-0.30	-0.29	-0.29
Temperature Coefficient (Voc)	V/°C	-0.173	-0.168	-0.169	-0.170	-0.172	-0.172
Temperature Coefficient (Isc)	mA/°C	1.10	0.85	0.86	0.87	0.88	0.88
Electrical Tolerance	%	+10/-5	+10/-5	+10/-5	+10/-5	+10/-5	+10/-5
Warranted Tolerance	%	+10/-0	+10/-0	+10/-0	+10/-0	+10/-0	+10/-0
Cell Efficiency	%	17.8	18.4	18.8	19.3	19.7	20.2
Module Efficiency	%	15.3	15.8	16.1	16.5	17.0	17.4
Power per Sq. Foot	W	14.2	14.7	15.0	15.4	15.8	16.2

Mechanical Specifications	
Internal Bypass Diodes	4 Bypass Diodes
Module Area (m²)	12.69 Ft² (1.18m²)
Weight (kg)	30.86 Lbs. (14kg)
Dimensions LxWxH (mm)	51.9x35.2x1.4in (1319x894x35mm)
Cable Length -Male/+Female (mm)	30.7/24.8in (780/630mm)
Cable Size / Connector Type	No.12 AWG / MC™ Connectors
Static Load Wind / Snow (Pa)	50PSF (2400Pa) / 39PSF (1876Pa)
Pallet Dimensions LxWxH (mm)	53x36x63in (1346x912x1600mm)
Pieces per Full Pallet / Weight (kg)	36pcs / 1111 Lbs (504kg)
Quantity per 20'/40'/53' Container	360pcs / 756pcs / 972pcs

Operating Conditions & Safety Ratings	
Temperature (°C)	-4°F to 104°F (-20°C to 40°C)[2]
Hail Safety Impact Velocity	1" hailstone (25mm) at 52mph (23m/s)
Fire Safety Classification	Class C
Safety & Rating Certifications	UL 1703, cUL, CEC
Limited Warranties	2 Years Workmanship / 20 Years Power Output

[1] STC Cell Temp 25°C, AM1.5, 1000W/m² [2] Monthly average low and high of the site

Dependence on Temperature
(Reference Data for 200 Watt Model)

Dependence on Irradiance
(Reference Data for 200 Watt Model)

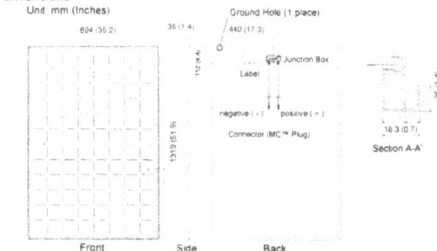

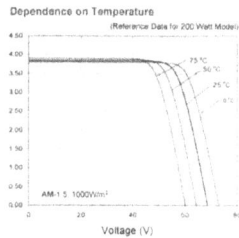

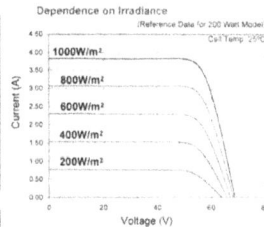

⚠ CAUTION! Read the operating instructions carefully before use of these products.

Note: Specifications and products above may change without notice. D4/01-07

Visit www.SANYO.com or contact an Authorized Representative for more information

Dimensions
Unit: mm (Inches)

Front   Side   Back

Figure B.1. Sanyo module specifications.

12. SPECIFICATIONS

· Under certain conditions, a photovoltaic module may produce more voltage and current than reported at Standard Test Conditions (STC). Refer to Section 690 of the National Electrical Code for guidance in series string sizing and choosing overcurrent protection.

Kyocera KDxxxGX-LPU Series Module Specification

Electrical Characteristics: @ STC			
Module Type	KD205GX-LPU	KD210GX-LPU	KD215GX-LPU
Pmax	205W (+5W/-0W)	210W (+5W/-0W)	215W (+5W/-0W)
Voc	33.2 V	33.2 V	33.2 V
Isc	8.36 A	8.58 A	8.78 A
Vpm	26.6 V	26.6 V	26.6 V
Ipm	7.71 A	7.90 A	8.09 A
Factory Installed Bypass Diode			
(QTY)		YES (3pcs)	
Series Fuse Rating (A)	15	15	15
Thermal Characteristics: Temp. Coefficient			
Voc (V/°C)	-1.20×10^{-1}	-1.20×10^{-1}	-1.20×10^{-1}
I sc (A/°C)	5.02×10^{-3}	5.15×10^{-3}	5.27×10^{-3}
Vpm (V/°C)	-1.38×10^{-1}	-1.39×10^{-1}	-1.39×10^{-1}
Physical Characteristics:			
Length		59.1" (1500 mm)	
Width		39.0" (990 mm)	
Depth		1.8" (46 mm)	
Weight		39.7 lb (18.0 kg)	
Mounting Hole		Diameter .35" (9mm) Quantity 4pcs	
Grounding Hole		Diameter .35" (9mm) Quantity 4pcs	
Application Class		Class A	

NOTES
(1) Standard Test Conditions of irradiance of 1000 W/m^2, spectrum of air mass 1.5, and cell temperature of 25 deg C.
(2) See module drawing for mounting and grounding holes locations.

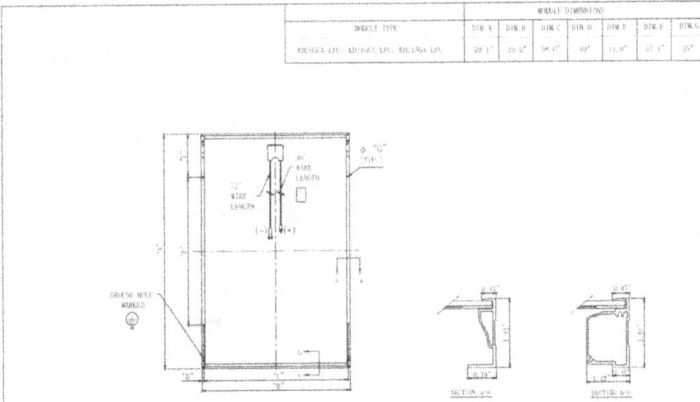

Figure B.2. Kyocera module specifications.

SOLON Blue 220/01 10/09 EN

SOLON Blue 220/01

Technical features:
- Highly efficient polycrystalline solar module
- Module efficiency of up to 14.33%
- Excellent low light response
- 0.16 in. (4 mm) solar glass and twin-wall frame profile for highest load capacity
- Optimal heat dissipation technology

SOLON advantages:
- 10 year product guarantee, 25 year performance guarantee
- Individual consulting and support from our service team and our international distribution partners
- On time delivery
- Certified quality products (UL listed)
- Extensive manufacturing and development experience

Electrical specifications – typical

Capacity rating* (±3%)	P$_{max}$	235 Wp	230 Wp	225 Wp	220 Wp
Module efficiency		14.33%	14.02%	13.72%	13.41%
Rated voltage	U$_{mpp}$	29.20 V	29.00 V	28.85 V	28.75 V
Rated current	I$_{mpp}$	8.05 A	7.95 A	7.80 A	7.65 A
Open circuit voltage	U$_{oc}$	36.9 V	36.65 V	36.55 V	36.40 V
Short circuit current	I$_{sc}$	8.65 A	8.55 A	8.40 A	8.30 A

The above values are effective for irradiation of 1,000 W/m², AM 1.5, and a cell temperature of 69.80°F (29°C) (standard test conditions). They are subject to production tolerances. The modules can be delivered with their characteristic data for the detailed system configuration.

Temperature coefficients

Tc of open circuit voltage	−0.34%/K
Tc of short circuit current	0.05%/K
Tc of power	−0.42%/K

Mechanical specification

Dimensions (H x W x D)	64.57 x 39.37 x 1.65 in. (1,640 x 1,000 x 42 mm)
Weight	51.81 lbs (23.5 kg)
Junction box	1 Tyco junction box with bypass diodes
Cable	Solar cable, length 39.37 in. (1,000 mm), 12 AWG (4 mm²), connectors prefabricated with Tyco plug
Front glass	Transparent toughened safety glass, 0.16 in. (4 mm)
Solar cells	60 cells polycrystalline Si 6.2", 6.14 x 6.14 in. (156 x 156 mm)
Cell encapsulation	EVA (Ethylene Vinyl Acetate)
Back side	White composite film
Frame	Anodized aluminium frame with twin-wall profile and drainage holes

Operating conditions

Temperature range	−40°F to +185°F (−40°C to +85°C)
Maximum system voltage	600 V
Maximum surface load capacity	Tested up to 113 psf (5,400 Pa) according to IEC 61215 (advanced test)
Resistance against hail	Maximum diameter of 1.1 in. (28 mm) with impact speed of 53.44 mph (86 km/h)

Guarantees and certifications

Product warranty	10 years
Performance guarantee	Guaranteed output of 90% for 10 years and 80% for 25 years
Approvals and certificates	UL listed, CEC registered

Drawing

39.37 ±0.08
64.57 ±0.08

Frame with drainage holes 1.65
12.99
4 holes for potential equalization Ø 0.18
12.99
0.53
Glass
SOLON frame profile
1.65
1.46
Dimensions in inches

(UL) ◻ C€

Subject to manufacturers. Electric data without guarantee

SOLON Corporation
6950 S. Country Club Road
Tucson · AZ · 85756-7151 · USA

Phone +1 520 807-1300
Fax +1 520 807-4046
E-Mail solon.us@solon.com

SOLON SE
Am Studio 16
12489 Berlin · Germany

Phone +49 30 81879-0
Fax +49 30 81879-9999
E-Mail components@solon.com

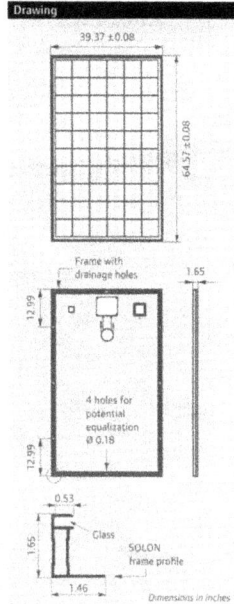

For more information on SOLON products please visit www.solon.com. We are glad to inform you personally as well.

Figure B.3. Solon module specifications.

PHOTOVOLTAIC

Technical Data

Electrical Data[1]

SCHOTT POLY™ 225/220/217/210

The electrical data applies to standard test conditions (STC): Irradiance at the module level of 1,000 W/m² with spectrum AM 1.5 and a cell temperature of 25°C.

		225 Wp	220 Wp	217 Wp	210 Wp
Nominal power	P_{nom}	225 Wp	220 Wp	217 Wp	210 Wp
Voltage at maximum power point	V_{mpp}	29.8 V	29.7 V	29.6 V	29.3 V
Current at maximum power point	I_{mpp}	7.55 A	7.41 A	7.33 A	7.16 A
Open circuit voltage	V_{oc}	36.7 V	36.5 V	36.4 V	36.1 V
Short circuit current	I_{sc}	8.24 A	8.15 A	8.10 A	7.95 A
Module efficiency	η_{mod}	14.80%	14.47%	14.28%	13.82%

Data at Nominal Operating Cell Temperature (NOCT)

Irradiance 800 W/m², spectrum Air Mass 1.5, windspeed 1m/s and a cell temperature of 20°C.

Nominal power	P_{nom}	161	158	156	151
Voltage at nominal power	V_{mpp}	26.9	26.7	26.7	26.4
Open circuit voltage	V_{oc}	33.5	33.3	33.2	33.0
Short circuit current	I_{sc}	6.60	6.53	6.49	6.37
Temperature (°C)	T_{NOCT}	47.2	47.2	47.2	47.2

Temperature Coefficients

Power	- 0.47 %/°C
Open circuit voltage	- .334 %/°C
Short circuit current	.030 %/°C

Characteristic Data

Solar cells per module	60
Cell type	MAIN-Isotextured (polycrystalline silicon) 6" (156 mm x 156 mm), full square
Connection	Junction box with 3 bypass diodes, PV WIRE, 43.3" x 4mm², TYCO SolarLok connectors
Front panel	Low-iron solar glass 4 mm
Frame material	Anodized aluminum

Dimensions and Weight

Dimensions	66.34" (1,685 mm) x 39.09" (993 mm) tolerance ± .118"
Thickness with frame	1.97" (50 mm) tolerance .04"
Weight	Approx. 50.6 lbs. (23.0 kg)

Limits

System Voltage (V_{DC})	600
Maximum Reverse Current I_R(A)*	15
Operating module temperature (°C)	-40...+85
Maximum load	75lbs/ft²
Fire Classification	C

No external current greater than V_{oc} shall be applied to the module.

Qualifications

The SCHOTT POLY™ 225/220/217/210 is certified to and meets the requirements of UL 1703.

[1]The rated power may vary by ± 4%.

The right is reserved to make technical modification. For detailed product drawings and specification please contact SCHOTT Solar or an authorized reseller.

SCHOTT Solar, Inc.

U.S. Sales and Marketing
Toll free: 888-457-6527
Email: solar.sales@us.schottsolar.com
www.us.schottsolar.com

U.S. Production Facility
5201 Hawking Drive, SE
Albuquerque, NM 87106
Phone: 505-212-8500

SCHOTT
solar

Figure B.4. Schott module specifications.

APPENDIX C

COLORADO NET METERING RULES

C.1 SUMMARY

Senate Bill 51 of April 2009 made several changes, effective September 1, 2009, to the net metering rules for investor-owned utilities as they apply to solar electric systems. These changes include shifting the maximum system size for solar electric systems from 2 MW to 120% of the annual consumption of the site; redefining a site to include all contiguous property owned by the consumer; and allowing system owners to make a one-time election in writing to have their annual net excess generation (NEG) carried forward as a credit from month to month indefinitely, rather than being paid annually at the average hourly incremental cost for that year. The Colorado Public Utilities Commission (PUC) incorporated these changes in the final rules they adopted in September 2009. While SB 51 dealt explicitly with solar electric systems, the final rules pertain to all eligible energy resources listed above.

Systems sized up to 120% of the customer's annual average consumption that generate electricity using qualifying renewable energy resources are eligible for net metering in investor owned utility (IOU) service territories. Municipal and cooperative utilities are subject to lesser capacity-based maximums as described below. Electricity generated at a customer's site can be applied toward meeting a utility's renewable-generation requirement under Colorado's renewable portfolio standard (RPS), though the renewable electricity certificates remain with the net metering customer unless purchased by the utility. The RPS mandates that 4% of the renewable requirement be met with solar energy; half of this percentage must come from solar electricity generated at customers' facilities.

Any customer's NEG in a given month is applied as a kilowatt-hour (kWh) credit to the customer's next bill. If, in a calendar year, a customer's generation exceeds consumption, the utility must reimburse the customer for the excess generation at the utility's average hourly incremental cost for the prior 12-month period. Net metering customers of an IOU may make a one-time election in writing on or before the end of the calendar year to have their NEG carried forward from month to month indefinitely. If customers choose this option, they will surrender all their kWh credits if and when they terminate service with their utility.

If a customer generator does not own a single bidirectional meter, then the utility must provide one free of charge. Systems over 10 kilowatts (kW) in capacity require a second meter to measure the output for the counting of renewable energy credits (RECs). Customers accepting IOU incentive payments must surrender all RECs for the next 20 y. Cooperative and municipal utilities are free to develop their own incentive programs at their discretion but they are not subject to the solar-specific requirements of the RPS.

House Bill 08-1160, enacted in March 2008, requires municipal utilities with more than 5,000 customers and all cooperative utilities to offer net metering. The new law allows residential systems up to 10 kW in capacity, and commercial and industrial systems up to 25 kW, to be credited monthly at the retail rate for any NEG their systems produce. Co-ops and municipal utilities are authorized to exceed these minimum size standards. The statute also requires the utilities to *pay* for any remaining NEG at the end of an annual period but does not define what the annual period is, nor the rate at which it will be paid. The law says that the utilities will make a payment based on a "rate deemed appropriate by the utility." The new law also required the PUC to open a new *rule making* to determine if the existing interconnection standards adopted in 4 CCR 723-3, Rule 3665, should be modified for co-ops. Municipal utilities are required to adopt rules "functionally similar" to the existing PUC rules, but may reduce or waive any of the insurance requirements.

C.2 BACKGROUND

In December 2005, the Colorado PUC first adopted standards for net metering and interconnection, as required by Amendment 37, a renewable energy ballot initiative approved by Colorado voters in November 2004.

REFERENCES

[1]. Messenger, R. A.; and J. Ventre. *Photovoltaic Systems Engineering*, 3. 3rd ed. Boca Raton, FL: CRC Press, 2003.

[2]. Patel, M. R. *Wind and Solar Power Systems: Design, Analysis, and Operation*, 3. 2nd ed. Boca Raton, FL: Taylor & Francis Group, 2005.

[3]. Stevenson, W. D. Jr. *Elements of Power System Analysis*, 4. 4th ed. McGraw Hill Higher Education, 1982.

[4]. International Energy Agency. *IEA Key Statistics 2010*-data. International Energy Agency, 2008. online at www.IEA.org

[5]. Freris, L.; and D. Infield. *Renewable Energy in Power Systems*, 2. UK: John Wiley & Sons, Ltd, 2008.

[6]. Stevenson, W. D. Jr. *Elements of Power System Analysis*, 15. 4th ed. USA: McGraw Hill Higher Education,1982.

[7]. Stevenson, W. D. Jr. *Elements of Power System Analysis*, 29. 4th ed. USA: McGraw Hill Higher Education, 1982.

[8]. Stevenson, W. D. Jr. *Elements of Power System Analysis*, 276. 4th ed. USA: McGraw Hill Higher Education, 1982.

[9]. Freris, L.; and D. Infield. *Renewable Energy in Power Systems*, 24. West Sussex, UK: John Wiley & Sons Ltd, 2008.

[10]. NREL. *Energy Efficiency & Renewable Energy*. US Department of Energy, NREL, April 2009.

[11]. Patel, M. R. *Wind and Solar Power Systems*, 17. 2nd ed. Washington, DC: CRP Press, 2005.

[12]. Messenger, R.A; and J. Ventre. *Photovoltaic Systems Engineering*, 15. 3rd ed. Boca Raton, FL: CRC Press, 2003.

[13]. Messenger, R. A.; and J. Ventre. *Photovoltaic System Engineering*, 27. 3rd ed. Boca Raton, FL: CRC Press, 2003

[14]. Solar Energy International. *Photovoltaics Design and Installation Manual*, 209. 1st ed. Gabriola Island, BC: New Society Publishers, September 1, 2004.

[15]. Solar Energy International. *Courtesy of Photovoltaics Design and Installation Manual*, 276. Carbondale, CO: Solar Energy International, 2004.

[16]. Honsberg, C.; et al. "Absorption Coefficient," *PVeducation.org*, November 30, 2012.

[17]. Messenger, R.A; and J. Ventre. *Photovoltaic System Engineering.* 3rd ed. Boca Raton, FL: CRC Press, 2003; Chapter 11.

[18]. Fthenakis, V.; and E. Alsema. "Photovoltaics Energy Payback Times, Greenhouse Gas Emissions and External Costs: 2004-Early 2005 Status," *Progress in Photovoltaics* 14, no. 3 (May 2006), pp. 275–280.

[19]. Solar Energy International. *Photovoltaics Design and Installation Manual,* 50. Carbondale, CO: Solar Energy International, 2006.

[20]. "NEC Article 690," *Solar Photovoltaic Systems.* Quincy, MA: National Fire Protection Association, 2008, pp. 70–578.

[21]. Patel, M. R. *Wind and Solar Power Systems,* 9. 2nd ed. Washington, DC: CRP Press, 2005.

[22]. Patel, M. R. *Wind and Solar Power Systems,* 48. 2nd ed. Washington, DC: CRP Press, 2005.

[23]. Price, G. D. "Wind Energy Analysis," *North Dakota Wind Generation Project,* 28–30. 1996.

[24]. Fitzgerald, A. E.; C. Kingsley Jr.; and S. D. Umans. *Electric Machinery,* 420. New York, NY: McGraw Hill Companies, Inc., 1983.

[25]. Woofenden, I.; and M. Sagrillo. "2010 Wind Generator-Buyer's Guide," *Home Power 137,* June and July 2010, p. 48.

[26]. Patel, M. R. *Wind and Solar Power Systems,* 242. 2nd ed. Washington, DC: CRP Press, 2006.

[27]. Unirac. *SunFrame Code-Compliant Installation Manual 809,* Table 2, p. 6. Albuquerque NM: Unirac—A Hill Group Company, 2013 (publications@ unirac.com).

GLOSSARY

absorbed glass mat (AGM): A fibrous silica glass mat to suspend the electrolyte in batteries. This mat provides pockets that assist in the recombination gasses generated during charging back into water.

alternating current (AC): Electric current in which the direction of flow is reversed at frequent intervals, usually 100 or 120 times per second (50 or 60 cycles/s or 50/60 Hz).

altitude: The angle between the horizon (a horizontal plane) and the sun's position in the sky, measured in degrees.

amorphous silicon: A noncrystalline semiconductor material that has no long-range order, often used in thin film photovoltaic modules.

ampere (A) or amp: The unit for the electric current; the flow of electrons. One amp is 1 coulomb passing in 1 s. One amp is produced by an electric force of 1 V acting across a resistance of 1 ohm. Sometimes this is abbreviated as I for intensity.

ampere-hour (Ah): Quantity of electrical energy equal to the flow of 1 A of current for 1 h. Typically used to quantify battery bank capacity.

angle of incidence: Angle that references the sun's radiation striking a surface. A normal angle of incidence refers to the sun striking a surface at a 90°angle.

array: Any number of photovoltaic modules connected together to provide a single electrical output at a specified voltage. Arrays are often designed to produce significant amounts of electricity.

autonomous system: A stand-alone PV system that has no back-up generating source. May or may not include storage batteries.

avoided cost: The minimum amount an electric utility is required to pay an independent power producer, under the PURPA regulations of 1978, equal to the costs the utility calculates it avoids in not having to produce that power (usually substantially less than the retail price charged by the utility for power it sells to customers).

azimuth: Angle between true south and the point directly below the location of the sun. Measured in degrees east or west of true south in northern latitudes.

balance of system (BOS): All system components and costs other than the PV modules. It includes design costs, land, site preparation, system installation, support structures, power conditioning, operation and maintenance costs, indirect storage, and related costs.

barrier energy: The energy given up by an electron in penetrating the cell barrier, a measure of the electrostatic potential of the barrier.

base power: Power generated by a utility unit that operates at a very high capacity factor.

baseline performance value: Initial values of I_{sc}, V_{oc}, P_{mp}, and I_{mp} measured by the accredited laboratory and corrected to standard test conditions, used to validate the manufacturer's performance measurements provided with the qualification modules as per IEEE 1262. See also *short-circuit current, open circuit voltage, maximum power point.*

battery: Two or more cells electrically connected for storing electrical energy. Common usage permits this designation to be applied also to a single cell used independently, as in a flashlight battery.

battery capacity: The total number of ampere-hours that can be withdrawn from a fully charged cell or battery.

battery cell: A galvanic cell for storage of electrical energy. This cell, after being discharged, may be restored to a fully charged condition by an electric current.

battery cycle life: The number of cycles, to a specified depth of discharge, that a cell or battery can undergo before failing to meet its specified capacity or efficiency performance criteria.

battery self-discharge: The rate at which a battery, without a load, will lose its charge.

battery state of charge: Percentage of full charge or 100% minus the depth of discharge.

blocking diode: A semiconductor device connected in series with a PV module and a storage battery to prevent a reverse current discharge of the battery through the module when there is no output, or low output from the cells. When connected in series to a PV string, it protects its modules from a reverse power flow preventing against the risk of thermal destruction of solar cells.

boron (B): A chemical element, atomic number 5, semimetallic in nature, used as a dopant to make p-semiconductor layers.

British thermal unit (BTU): The amount of heat energy required to raise the temperature of 1 lb. of water from 60°F to 61°F at 1 atmosphere pressure. Roughly equivalent to the amount of energy released by burning 1 match stick.

building-integrated photovoltaics (BIPV): A term for the design and integration of PV into the building envelope, typically replacing conventional building materials. This integration may be in vertical facades, replacing view glass, spandrel glass, or other facade material; into semitransparent skylight systems; into roofing systems, replacing traditional roofing materials; into shading eyebrows over windows; or other building envelope systems.

bypass diode: A diode connected across one or more solar cells in a photovoltaic module so that the diode will conduct if the cell(s) become reverse biased. Alternatively, a diode connected antiparallel across a part of the solar cells of a PV module. It protects these solar cells from thermal destruction in case of total or partial shading of individual solar cells while other cells are exposed to full light.

cadmium (Cd): A chemical element, atomic number 48, used in making certain types of solar cells and batteries.

cadmium telluride (CdTe): A polycrystalline, thin-film photovoltaic material.

capacity factor: The amount of energy that the system produces at a particular site as a percentage of the total amount that it would produce if it operated at rated capacity during the entire year. For example, the capacity factor for a wind farm ranges from 20% to 35%.

cathodic protection: A method of preventing oxidation (rusting) of exposed metal structures, such as bridges and pipelines, by imposing between the structure and the ground a small electrical voltage that opposes the flow of electrons and that is greater than the voltage present during oxidation.

cell: The basic unit of a photovoltaic module. This word is also commonly used to describe the basic unit of batteries (i.e., a 6 V battery has three 2 V cells).

cell barrier: A very thin region of static electric charge along the interface of the positive and negative layers in a photovoltaic cell. The barrier inhibits the movement of electrons from one layer to the other, so that higher energy electrons from one side diffuse preferentially through it in one direction, creating a current and thus a voltage across the cell. Also called depletion zone, cell junction, or space charge.

cell junction: The area of immediate contact between two layers (positive and negative) of a photovoltaic cell. The junction lies at the center of the cell barrier or depletion zone.

central power: The generation of electricity in large power plants with distribution through a network of transmission lines (grid) for sale to a number of users. Opposite of distributed power.

charge controller: A device that controls the charging rate and/or state of charge for batteries.

charge rate: The current applied to a cell or battery to restore its available capacity.

chemical vapor deposition (CVD): A method of depositing thin semiconductor films. With this method, a substrate is exposed to one or more vaporized compounds, one or more of which contain desirable constituents. A chemical reaction is initiated, at or near the substrate surface, to produce the desired material that will condense on the substrate.

cleavage of lateral epitaxial films for transfer (CLEFT): A process for making inexpensive GaAs photovoltaic cells in which a thin film of GaAs is grown atop a thick, single-crystal GaAs (or other suitable material) substrate and then is cleaved from the substrate and incorporated into a cell, allowing the substrate to be reused to grow more thin-film GaAs.

coal: A black, solid fossil fuel, usually found underground. Coal is often burned to make electricity in utility-scale production.

combined collector: A photovoltaic device or module that provides useful heat energy in addition to electricity.

compact fluorescent lights: Lights that use a lot less energy than regular light bulbs. We can use compact fluorescent lights for reading lights and ceiling lights.

concentrator: A PV module that uses optical elements to increase the amount of sunlight incident on a PV cell. Concentrating arrays must track the sun and use only the direct sunlight because the diffuse portion cannot be focused onto the PV cells.

conversion efficiency: The ratio of the electric energy produced by a photovoltaic device (under full sun conditions) to the energy from sunlight incident upon the cell.

copper indium diselenide (CuInSe2 or CIS): A polycrystalline thin-film photovoltaic material (sometimes incorporating gallium (CIGS) and/or sulfur).

crystalline silicon: A type of PV cell made from a single crystal or polycrystalline slice of silicon.

current: The flow of electric charge in a conductor between two points having a difference in potential (voltage).

current at maximum power (I_{mp}): The current at which maximum power is available from a module. Refer to standard 1703, Underwriters Laboratories [UL 1703].

cycle life: Number of discharge–charge cycles that a battery can tolerate under specified conditions before it fails to meet specified criteria as to performance (e.g., capacity decreases to 80% of the nominal capacity).

Czochralski process: A method of growing large, high-quality semiconductor crystal by slowly lifting a seed crystal from a molten bath of the material under careful cooling conditions.

days of autonomy: The number of consecutive days a stand-alone system battery bank will meet a defined load without solar energy input.

DC to DC converter: Electronic circuit to convert DC voltages (e.g., PV module voltage) into other levels (e.g., load voltage). Can be part of a maximum power point tracker (MPPT).

deep cycle battery: Type of battery that can be discharged to a large fraction of capacity many times without damaging the battery.

deep discharge: Discharging a battery to 50% or less of its full charge.

depth of discharge (DOD): The amount of ampere-hours removed from a fully charged cell or battery, expressed as a percentage of rated capacity.

design month: The month having the combination of insolation and load that requires the maximum energy from the array.

diffuse insolation: Sunlight received indirectly as a result of scattering due to clouds, fog, haze, dust, or other obstructions in the atmosphere. Opposite of direct insolation.

diode: Electronic component that allows current flow in one direction only.

direct current (DC): Electric current in which electrons flow in one direction only. Opposite of alternating current.

direct insolation: Full sunlight falling directly upon a collector. Opposite of diffuse insolation.

discharge rate: The rate, usually expressed in amperes over time, at which electrical current is taken from the battery.

disconnect: Switch gear used to connect or disconnect components of a PV system for safety or service.

distributed power: Generic term for any power supply located near the point where the power is used. Opposite of central power. See also *stand-alone* and *remote site*.

dopant: A chemical element (impurity) added in small amounts to an otherwise pure semiconductor material to modify the electrical properties of the material. An n-dopant introduces more electrons. A p-dopant creates electron vacancies (holes).

doping: The addition of dopants to a semiconductor.

duty cycle: The ratio of active time to total time. Used to describe the operating regime of appliances or loads.

edge-defined film-fed growth (EFG): A method for making sheets of polycrystalline silicon in which molten silicon is drawn upward by capillary action through a mold.

efficiency: The ratio of output power to input power. Expressed as a percentage.

electric circuit: Path followed by electrons from a power source (generator or battery) through an external line (including devices that use the electricity) and returning through another line to the source.

electric current: A flow of electrons; electricity.

electrical grid: An integrated system of electricity distribution, usually covering a large area.

electrodeposition: Electrolytic process in which a metal is deposited at the cathode from a solution of its ions.

electrolyte: A liquid conductor of electricity in which flow of current takes place by migration of ions. The electrolyte for a lead–acid storage cell is an aqueous solution of sulfuric acid.

energy: The ability to do work. Stored energy becomes working energy when we use it.

energy audit: A survey that shows how much energy you use in your house, apartment, or business. It can indicate your most intensive energy-consuming appliances and even identify heating and cooling leaks that will help you find ways to use less energy.

energy density: The ratio of energy available from a battery to its volume (Wh/1) or mass (Wh/kg).

energy payback time: The time required for any energy-producing system or device to produce as much energy as was required in its manufacture.

equalization: The process of mixing the electrolyte in batteries by periodically overcharging the batteries for a short period to refresh cell capacity.

fill factor: The ratio of a photovoltaic cell's actual power to its power if both current and voltage were at their maxima. A key characteristic in evaluating cell performance.

flat-plate PV: Refers to a PV array or module that consists of nonconcentrating elements. Flat-plate arrays and modules use direct and diffuse sunlight, but if the array is fixed in position, some portion of the direct sunlight is lost because of oblique sun-angles in relation to the array.

float charge: Float charge is the voltage required to counteract the self-discharge of the battery at a certain temperature.

float life: Number of years that a battery can keep its stated capacity when it is kept at float charge (see *float charge*).

fossil fuels: Fuels formed in the ground from the remains of dead plants and animals. It takes millions of years to form fossil fuels. Oil, natural gas, and coal are fossil fuels.

fuel: Any material that can be burned to make energy.

gassing current: Portion of charge current that goes into electrolytical production of hydrogen and oxygen from the electrolytic liquid in the battery. This current increases with increasing voltage and temperature.

gel-type battery: Lead–acid battery in which the electrolyte is composed of a silica gel matrix.

gigawatt (GW): One billion watts. One million kilowatts. One thousand megawatts.

glazings: Clear materials (such as glass or plastic) that allow sunlight to pass into solar collectors and solar buildings, trapping heat inside.

grain boundaries: The boundaries where crystallites in a multicrystalline material meet.

grid: See *electrical grid*.

grid-connected: A PV system in which the PV array acts like a central generating plant, supplying power to the grid.

grid-interactive: See *grid-connected*.

hybrid system: A PV system that includes other sources of electricity generation, such as wind or fossil fuel generators.

I–V curve: A graphical presentation of the current versus the voltage from a photovoltaic device as the load is increased from the short circuit (no load) condition to the open circuit (maximum voltage) condition.

incident light: Light that shines onto the surface of a solar cell or module.

infrared radiation: Electromagnetic radiation whose wavelengths lie in the range from 0.75 to 1,000 μm.

insolation: Sunlight, direct or diffuse; [derived from "incident solar radiation"]; usually expressed in watts per square meter [not to be confused with *insulation*].

insulation: Materials that reduce the rate or slow down the movement of heat.

interconnect: A conductor within a module or other means of connection that provides an electrical interconnection between the solar cells.

inverter: A device that converts DC electricity into AC electricity (single or multiphase), either for stand-alone systems (not connected to the grid) or for utility-interactive systems.

junction box: An electrical box designed to be a safe enclosure in which to make proper electrical connections. On PV modules, this is where PV strings are electrically connected.

kilowatt (kW): 1,000 W.

kilowatt-hour (kWh): One thousand watt hours. The kWh is a unit of energy. 1 kWh = 3,600 kilo-joules (kJ).

life cycle cost: An estimate of the cost of owning and operating a system for the period of its useful life; usually expressed in terms of the present value of all lifetime costs.

line-commutated inverter: An inverter that is tied to a power grid or line. The commutation of power (conversion from DC to AC) is controlled by the power line so that, if there is a failure in the power grid, the PV system cannot feed power into the line. Also called grid-connected (GC) inverter.

load: Anything in an electrical circuit that, when the circuit is turned on, draws power from that circuit.

maximum power point (MPP): The point on the current–voltage (I–V) curve of a module under illumination, where the product of current and voltage is maximum. For a typical silicon cell, MPP occurs at about 0.45 V.

maximum power point tracker (MPPT): Means of a power conditioning unit that automatically operates the PV generator at its MPP under all conditions.

megawatt (MW): One million watts; 1,000 kW.

module: See *photovoltaic module*.

multicrystalline: Material that is solidified at such as rate that many small crystals (crystallites) form. The atoms within a single crystallite are symmetrically arranged, whereas crystallites are jumbled together. These numerous grain boundaries reduce the device efficiency. A material composed of variously oriented, small individual crystals. (sometimes referred to as polycrystalline or semicrystalline).

n-type semiconductor: A semiconductor produced by doping an intrinsic semiconductor with an electron-donor impurity, for example, phosphorus in silicon.

NEC: An abbreviation for the National Electrical Code®, which contains safety guidelines and required practices for all types of electrical installations. Articles 690 and 694 pertain to solar photovoltaic systems and small wind generator systems.

nominal operating cell temperature (NOCT): The reference cell (module) operating temperature presented on manufacturer's literature. Generally, the NOCT is referenced at 25°C, 77°F.

nominal voltage: A reference voltage used to describe batteries, modules, or systems (e.g., a 12-, 24-, or 48-V battery, module, or system).

nonrenewable fuels: Fuels that cannot be easily made or renewed. We can use up nonrenewable fuels. Oil, natural gas, and coal are nonrenewable fuels.

ohm: The unit of resistance to the flow of an electric current.

one-axis tracking: A system capable of rotating about one axis, also referred to as single axis. These tracking systems usually follow the sun from east to west throughout the day.

open-circuit voltage (VOC): The maximum possible voltage across a photovoltaic cell or module; the voltage across the cell in sunlight when no current is flowing.

orientation: Placement according to the compass directions: north, south, east, and west.

p-type silicon: Semiconductor-grade silicon doped with the element boron giving it a positive bias.

p–n: A semiconductor device structure in which the junction is formed between a p-type layer and an n-type layer.

panel: See *photovoltaic panel.*

parallel connection: A way of joining two or more electricity-producing devices such as PV cells or modules, or batteries by connecting positive leads together and negative leads together; such a configuration increases the current but the voltage is constant.

passive solar building: A building that utilizes nonmechanical, nonelectrical methods for heating, cooling, and/or lighting.

peak load; peak demand: The maximum load, or usage, of electrical power occurring in a given period of time, typically a day.

peak power: Power generated by a utility unit that operates at a very low capacity factor; generally used to meet short-lived and variable high-demand periods.

peak sun hours (PSH): The equivalent number of hours per day when solar irradiance averages $1,000$ W/m^2 (full sun).

phosphorus (P): A chemical element, atomic number 15, used as a dopant in making n-semiconductor layers.

photon: A particle of light that acts as an individual unit of energy.

photovoltaic (PV): Pertaining to the direct conversion of photons of sunlight into electricity.

photovoltaic array: An interconnected system of PV modules that function as a single electricity-producing unit. The modules are assembled as a discrete structure, with common support or mounting. In smaller systems, an array can consist of a single module.

photovoltaic cell: The smallest semiconductor element within a PV module to perform the immediate conversion of light into electrical energy (DC voltage and current).

photovoltaic conversion efficiency: The ratio of the electric power produced by a photovoltaic device to the power of the sunlight incident on the device.

photovoltaic module: The smallest environmentally protected, essentially planar assembly of solar cells and ancillary parts, intended to generate DC power under unconcentrated sunlight. The structural (load carrying) member of a module can either be the top layer (superstrate) or the back layer (substrate).

photovoltaic panel: Often used interchangeably with *photovoltaic module* (especially in one-module systems), but more accurately used to refer to a physically connected collection of modules (i.e., a laminate string of modules used to achieve a required voltage and current).

photovoltaic peak watt: Maximum rated output of a cell, module, or system. Typical rating conditions are $1,000$ W/m^2 of sunlight, 68°F (20°C) ambient air temperature and 1 m/s wind speed.

photovoltaic system: A complete set of components for converting sunlight into electricity by the photovoltaic process, including the array and balance of system components.

physical vapor deposition: A method of depositing thin semiconductor films. With this method, physical processes, such as thermal evaporation or bombardment of ions, are used to deposit an elemental semiconductor material on a substrate.

polycrystalline: See *multicrystalline*.

power conditioning equipment: Electrical equipment, or power electronics, used to convert power from a photovoltaic array or wind turbine generator into a form suitable for subsequent residential or commercial use. A collective term for inverter, converter, battery charge regulator, or blocking diode.

power factor: The ratio of the average power and the apparent volt-amperes.

pulse-width-modulated wave inverter (PWM): PWM inverters are the most expensive, but produce a high quality of output signal at minimum current harmonics. The output voltage is very close to sinusoidal.

quad: A measure of energy equal to 1 trillion BTUs; an energy equivalent to approximately 172 million barrels of oil.

qualification test: A procedure applied to a selected set of PV modules involving the application of defined electrical, mechanical, or thermal stress in a prescribed manner and amount. Test results are subject to a list of defined requirements.

rectifier: A device that converts AC to DC. See *inverter*.

remote site: Site that is not located near the utility grid.

resistance (R): The property of a conductor that opposes the flow of an electric current, resulting in the generation of heat in the conducting material. The unit of resistance is an *ohm*.

satellite power system (SPS): Concept for providing large amounts of electricity for use on the Earth from one or more satellites in geosynchronous Earth orbit. A very large array of solar cells on each satellite would provide electricity, which would be converted to microwave energy and beamed to a receiving antenna on the ground. There, it would be reconverted into electricity and distributed the same as any other centrally generated power, through a grid.

semiconductor: Any material that has a limited capacity for conducting an electric current. Certain semiconductors, including silicon, gallium arsenide, copper indium diselenide, and cadmium telluride, are uniquely suited to the photovoltaic conversion process.

semicrystalline: See *multicrystalline*.

series connection: A way of joining electrical equipment by connecting positive leads to negative leads; such a configuration increases the voltage while current remains the same.

series regulator: Type of battery charge regulator where the charging current is controlled by a switch connected in series with the PV module or array.

shelf life of batteries: The length of time, under specified conditions, that a battery can be stored so that it keeps its guaranteed capacity.

short-circuit current (I_{sc}): The current flowing freely from a photovoltaic cell through an external circuit that has no load or resistance; the maximum current possible.

shunt regulator: Type of a battery charge regulator where the charging current is controlled by a switch connected in parallel with the PV generator. Overcharging of the battery is prevented by shorting the PV generator.

silicon (Si): A chemical element, atomic number 14, semimetallic in nature, dark gray, and an excellent semiconductor material. A common constituent of sand and quartz (as the oxide). Crystallizes in face-centered cubic lattice, like a diamond. The most common semiconductor material used in making photovoltaic devices.

sine wave inverter: An inverter that produces utility-quality, sine wave power forms.

single-crystal material: A material that is composed of a single crystal or a few large crystals.

solar cell: See *photovoltaic cell.*

solar constant: The strength of sunlight; 1,353 W/m² in space and about 1,000 W/m² at sea level at the equator at solar noon.

solar energy: Energy from the sun. For example, the heat that builds up in your car when the windows are closed is solar energy.

solar noon: That moment of the day that divides the daylight hours for that day exactly in half. To determine solar noon, calculate the length of the day from the time of sunset and sunrise and divide by 2. The moment the sun is highest in the sky.

solar spectrum: The total distribution of electromagnetic radiation emanating from the sun.

solar thermal electric: Method of producing electricity from solar energy by using focused sunlight to heat a working fluid, which in turn drives a turbogenerator.

solar-grade silicon: Intermediate-grade silicon used in the manufacture of solar cells. Less expensive than electronic-grade silicon.

square wave inverter: The inverter consists of a DC source, four switches, and the load. The switches are power semiconductors that can carry a large current and withstand a high voltage rating. The switches are turned on and off at a correct sequence, at a certain frequency. The square wave inverter is the simplest and the least expensive to purchase, but it produces the lowest quality of power.

Staebler–Wronski effect: The tendency of amorphous silicon photovoltaic devices to lose efficiency upon initial exposure to light; named after Dr. David Staebler and Dr. Christopher Wronski.

stand-alone: An autonomous or hybrid photovoltaic system not connected to a grid. Some stand-alone systems require batteries or some other form of storage.

stand-off mounting: Technique for mounting a PV array on a sloped roof, which involves mounting the modules a short distance above the pitched roof. This promotes air flow to cool the modules.

Standard test conditions (STC): Conditions under which a module is typically tested in a laboratory: (a) Irradiance intensity of 1,000 W/m^2 (0.645 W/in^2), (b) AM1.5 solar reference spectrum, and (c) a cell (module) temperature of 25°C, plus or minus 2°C (77°F, plus or minus 3.6°F).

state of charge (SOC): The available capacity remaining in a cell or battery, expressed as a percentage of the rated capacity. For example, if 25 Ah have been removed from a fully charged 100 Ah cell, the state of charge is 75%.

substrate: The physical material upon which a photovoltaic cell is made.

sulfation: A condition that afflicts unused and discharged batteries; large crystals of lead sulfate grow on the plate, instead of the usual tiny crystals, making the battery extremely difficult to recharge.

superconductivity: The pairing of electrons in certain materials that, when cooled below a critical temperature, cause the material to lose all resistance to electricity flow. Superconductors can carry electric current without any energy losses.

superstrate: The covering on the sun side of a PV module, providing protection for the PV materials from impact and environmental degradation while allowing maximum transmission of the appropriate wavelengths of the solar spectrum.

surge: The momentary start-up condition of a motor requiring a large amount of electrical current.

surge capacity: The ability of an inverter or generator to deliver high currents momentarily required when starting a motor.

temperature compensation: An allowance made in charge controller set points for changing battery temperatures.

thermal electric: Electric energy derived from heat energy, usually by heating a working fluid, which drives a turbogenerator. See *solar thermal electric*.

thermal mass: Materials, typically masonry, that store heat in a passive solar home.

thin film: A layer of semiconductor material, such as copper indium diselenide, cadmium telluride, gallium arsenide, or amorphous silicon; a few microns or less in thickness; used to make photovoltaic cells.

tilt angle: Angle of inclination of collector as measured in degrees from the horizontal. For maximum performance, solar collectors/modules should be set at an angle perpendicular to the sun.

total harmonic distortion (THD): The measure of closeness in shape between a waveform and its fundamental component.

tracking PV array: PV array that follows the path of the sun to maximize the solar radiation incident on the PV surface. The two most common orientations are (a) one-axis tracking, where the array tracks the sun east to west and (b) two-axis tracking, where the array points directly at the sun at all times. Tracking arrays use both the direct and diffuse sunlight. Two-axis tracking arrays capture the maximum possible daily energy.

transformer: An electromagnetic device used to convert AC electricity, either to increase or decrease the voltage.

transmission lines: Conductors and structures used to transmit high-voltage electricity from the source to the electric distribution system.

trickle charge: A charge at a low rate, balancing through self-discharge losses, to maintain a cell or battery in a fully charged condition.

two-axis tracking: A system capable of rotating independently about two axes and following the sun's orientation and height in the sky (e.g., vertical and horizontal).

ultraviolet (UV): Electromagnetic radiation in the wavelength range of 4 to 400 nanometers.

uninterruptible power supply (UPS): The designation of a power supply providing continuous uninterruptible service when a main power source is lost.

utility-interactive inverter: An inverter that can function only when tied to the utility grid, and uses the prevailing line-voltage frequency on the utility line as a control parameter to ensure that the PV system's output is fully synchronized with the utility power.

vacuum deposition: Method of depositing thin coatings of a substance by heating it in a vacuum system.

vacuum evaporation: The deposition of thin films of a semiconductor material by the evaporation of elemental sources in a vacuum.

volt (V): A unit of measure of the force, or push, given the electrons in an electric circuit. One volt produces 1 A of current when acting against a resistance of 1 ohm.

voltage at maximum power (V_{mp}): The voltage at which maximum power is available from a module.

wafer: A thin sheet of semiconductor material made by mechanically sawing it from a single-crystal or multicrystal ingot or casting.

watt (W): The unit of electric power, or amount of work. One ampere of current flowing at a potential of 1 V produces 1 W of power.

watt-hour (Wh): A quantity of electrical energy when 1 W is used for 1 h.

waveform: The shape of the curve graphically representing the change in the AC signal voltage and current amplitude, with respect to time.

ACRONYMS AND ABBREVIATIONS

AC	Alternating current
Ah	Amp hours
AMI	Air mass index
ASES	American Solar Energy Society
AWEA	American Wind Energy Association
BTU	British thermal unit
Cp	Rotor energy conversion efficiency
DC	Direct current
DOE	Department of energy
EPRI	Electric Power Research Institute
GC	Grid Connected
GSHP	Ground Source Heat Pump
GWh	Giga watt hours
GW	Giga watts
HVDC	High voltage direct current
IEEE	Institute of Electrical and Electronics Engineers
ISES	International Solar Energy Society
IOU	Investor owned utility
kW	Kilowatt
kWh	Kilowatt hours
LC	Line commutated
MPP	Maximum power point
MPPT	Maximum power point tracking
MW	Megawatt
MWh	Mega-watt hours
NEC	National electric code
NEG	Net excess generation
NOAA	National Oceanic and Atmospheric Administration
NREL	National Renewable Energy Laboratory
NWTC	National Wind Test Center
PBI	Production based incentive
PURPA	Public Utility Regulatory Policies Act
PV	Photovoltaic

PWM	Pulse width modulation
QF	Qualifying facility
RE	Renewable energy
REC	Renewable energy credit
SOC	State of charge
SRC	Specific rated capacity (kW/m^2)
STC	Standard test conditions
THD	Total harmonic distortion
THM	Total head mass (nacelle + rotor)
TSR	Tip speed ratio (rotor)
UL	Underwriter Laboratories
VOC	Open circuit voltage

INDEX

THIS TITLE IS FROM OUR POWER GENERATION COLLECTION. OTHER COLLECTIONS INCLUDE...

Industrial Engineering

- Industrial, Systems, and Innovation Engineering — William R. Peterson, Collection Editor
- Manufacturing and Processes — Wayne Hung, Collection Editor
- Manufacturing Design
- General Engineering — Dr. John K. Estell and Dr. Kenneth J. Reid, Collection Editors

Electrical Engineering

- Electrical Power — Hemchandra M. Shertukde, Ph.D., P.E., Collection Editor
- Communications and Signal Processing — Orlando Baiocchi, Collection Editor
- Computer Engineering — Augustus (Gus) Kinzel Uht, PhD, PE, Collection Editor
- Electronic Circuits and Semiconductor Devices — Ashok Goel, Collection Editor

Civil and Environmental Engineering

- Environmental Engineering — Francis Hopcroft, Collection Editor
- Geotechnical Engineering — Dr. Hiroshan Hettiarachchi, Collection Editor
- Transportation Engineering — Dr. Bryan Katz, Collection Editor
- Sustainable Systems Engineering — Dr. Mohammad Noori, Collection Editor

Material Science

- Materials Characterization and Analysis — Dr. Richard Brundle, Collection Editor
- Mechanics & Properties of Materials
- Computational Materials Science
- Biomaterials

Not only is Momentum Press actively seeking collection editors for Collections, but the editors are also looking for authors. For more information about becoming an MP author, please go to http://www.momentumpress.net/contact!

Announcing Digital Content Crafted by Librarians

Momentum Press offers digital content as authoritative treatments of advanced engineering topics, by leaders in their fields. Hosted on ebrary, MP provides practitioners, researchers, faculty and students in engineering, science and industry with innovative electronic content in sensors and controls engineering, advanced energy engineering, manufacturing, and materials science. **Momentum Press offers library-friendly terms:**

- perpetual access for a one-time fee
- no subscriptions or access fees required
- unlimited concurrent usage permitted
- downloadable PDFs provided
- free MARC records included
- free trials

The **Momentum Press** digital library is very affordable, with no obligation to buy in future years.

For more information, please visit **www.momentumpress.net/library** or to set up a trial in the US, please contact **mpsales@globalepress.com**.

www.ingramcontent.com/pod-product-compliance
Lightning Source LLC
Chambersburg PA
CBHW070719220326
41598CB00024BA/3235